MATHS SKILLS *for* Advanced Sciences

Ken Price

OXFORD
UNIVERSITY PRESS

Oxford University Press, Great Clarendon Street, Oxford OX2 6DP

Oxford University Press is a department of the University of Oxford
it furthers the University's objective of excellence in research, scholarship,
and education by publishing worldwide in

Oxford New York

*Athens Auckland Bangkok Bogotá Buenos Aires Calcutta
Cape Town Chennai Dar es Salaam Delhi Florence Hong Kong Istanbul
Karachi Kuala Lumpar Madrid Melbourne Mexico City Mumbai
Nairobi Paris São Paulo Singapore Taipei Tokyo Toronto Warsaw*

with associated companies in *Berlin Ibadan*

Oxford is a registered trade mark of Oxford University Press
in the UK and in certain other countries

First published 2000

British Library Cataloguing in Publication Data

Data available

ISBN 0 19 914740 X

Typesetting, design and illustration by Hardlines, Charlbury, Oxford
Printed in Great Britain

Introduction

Many students following advanced science courses find their enjoyment of these courses and their progress hindered by inadequate mathematical skills. The application of previously learned skills set in a scientific context, and the need for more advanced skills, both make science less accessible. This book is intended to provide a reference of mathematical skills to support such students and to provide a useful checklist of skills for others.

- Each skill is covered in one or two pages of text
- Where skills covered in other sections are helpful, these are identified
- The treatment of each section includes:
  ✓ the basic principles
  ✓ examples of the use of the principles in typical advanced level situations
  ✓ useful hints to avoid errors
  ✓ practice questions (in most sections)
  ✓ answers to the questions
- To avoid any confusion the treatment focuses on the use of the skills rather than any background to the mathematical techniques

HOW TO USE THIS BOOK

Students:

- the book is is designed as a toolbox of mathematical skills to help you with your advanced science courses
- you can use it by working through the book in a structured way: to reinforce skills or learn new skills
- use the contents page to identify the skills you need in your biology, chemistry, physics, or technology courses (There will be some variations between syllabuses.)
- use the book as a toolbox, dipping into the appropriate pages when you need to learn or clarify a particular skill
- practise the skills using the examples provided

Teachers Note:

- *The book provides a useful basis for a teacher-led course to support those students who are not taking a mathematics course as a component of their advanced studies*
- *Teachers could use the book as a scheme of work and progress through it in a structured way*
- *Teachers could direct students to appropriate sections of the book having identified the needs of individuals, taking into account their abilities and areas of study*

CONTENTS

The columns BCPT indicate that the material on this page will be useful to students who are studying biology, chemistry, physics, and some technology courses. 'Tick boxes' are included for you to mark pages that are particularly relevant.

Symbols in common use

MATHEMATICAL SYMBOLS

The main symbols you will come across in science at this level are the familiar arithmetical symbols +, –, x and ÷. There are a number of others, however, which you should know and be able to use.

Symbol	Meaning	Example of its use
±	plus or minus	This is used in defining the uncertainty of a quantity. When a number is written as, for example, 5.6 ± 0.2 m, it means that the value lies in the range (5.6 – 0.2) m to (5.6 + 0.2) m, i.e. between 5.4 and 5.8 m.
Δ	a change in	Velocity $= \frac{\Delta s}{\Delta t}$ means that a change in distance of Δs occurs when time changes by Δt. It is used in writing the rate of change of one quantity with respect to another.
≈	approximately equal to	This is used when answers have been rounded off, e.g. when estimating or giving orders of magnitude. For example, the length of a virus ≈ 10^{-7}m.
∝	is directly proportional to	Force ∝ extension ($F \propto x$) means that F = a constant multiplied by x. When x is doubled F is doubled, when x is trebled F is trebled, etc. Frequency $\propto \frac{1}{\text{length}}$ means that frequency is *directly* proportional to $\frac{1}{\text{length}}$. It may also be stated as frequency is *inversely* proportional to length.
>	is greater than	Length >1.5 m means that the length is always greater than 1.5 m. The length is never equal to 1.5 m or less than 1.5 m.
<	is less than	This is used in a similar way to >.
>>	is much greater than	The speed of light is much greater than the speed of an aeroplane ($c >> v$). If ($c + v$) or ($c - v$) appeared in a calculation the v could be ignored: both of these are close to c.
<<	is much less than	This is used in a similar way to >>.
Σ	sum of	This is often seen in statistical analysis. Σx means the sum of all the quantities of x in a list. If the values for x are 2.5 cm, 3.5 cm and 4.0 cm, Σx = 10.0 cm.
≤	is less than or equal to	'The ants in the sample were ≤ 6.0 mm in length' would mean that there was at least one ant of length 6.0 mm and all others were smaller than this.
≥	is greater than or equal to	This is used in a similar way to ≤.
$\bar{x}$ or $<x>$	average or mean of all the values *of x*	This is frequently used in statistical analysis of data. In physics the mean square speed of molecules in a gas is $<c^2>$. This means that all the speeds are first squared and then the mean (average) value is found.

GREEK LETTERS

These are often used for physical quantities in equations. They are also used for naming subatomic particles. You need to know what they stand for so that you can communicate equations orally. You will also need to write down equations and symbols for particles. The whole Greek alphabet is listed below. The letters in bold are the ones you are likely to meet in advanced-level science. Where the capital form of the letter is also used, this is included in brackets.

α	**alpha**	ν	**nu**
β	**beta**	ξ	xi
γ	**gamma**	ο	omicron
δ (Δ)	**delta**	π	**pi**
ε	**epsilon**	ρ	**rho**
ζ	zeta	σ (Σ)	**sigma**
η	**eta**	τ	**tau**
θ (Θ)	**theta**	υ	upsilon
ι	iota	φ (Φ)	**phi**
κ	kappa	χ	**chi**
λ	**lambda**	ψ	**psi**
μ	**mu**	ω (Ω)	**omega**

Note: Symbols in textbooks

In textbooks, symbols that represent physical quantities are printed in italics and prefixes and units are written in upright type. For example:

$F = ma$

- F is measured in N
- m is in kg
- a is in m s^{-2}

Indices

Many scientific laws such as Newton's gravitational law ($F = GMm/r^2$) involve indices (singular = index). Indices are also used in numerical work when using standard form (also called scientific notation) and when writing down units for quantities, so you need to know how to handle them.

Positive indices

A positive index is a shorthand way of writing how many times a quantity is multiplied by itself, e.g.

$2.5^3 = 2.5 \times 2.5 \times 2.5 \approx 16$

$10^2 = 10 \times 10 = 100$

r^n means r multiplied by itself until n of them have been multiplied together.

What is an index?

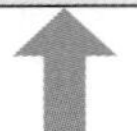

r^n

n is the index.

r is called the base.

We say that r is raised to the power of n.

Power and **exponent** are other names for an index.

Fractional or decimal indices

Important fractional indices are $\frac{1}{2}$ and $\frac{1}{3}$. In decimal form these would be 0.5 and 0.33.

Writing $T^{1/2}$ is the same as writing the square root of T ($\sqrt{T}$).

$r^{1/3}$ is the same as the cube root of r, i.e. the number which gives r when three of them are multiplied together.

$8^{1/3} = 2$, since $2 \times 2 \times 2 = 8$

Negative indices

The negative sign in an index shows that it is a **reciprocal**.

Writing r^{-1} is the same as writing $\frac{1}{r}$.

10^{-2} is the same as $\frac{1}{10^2}$

which is $\frac{1}{(10 \times 10)}$ or $\frac{1}{100}$ or 0.01

10^{-4} is the same as $\frac{1}{10^4}$

which is $\frac{1}{(10 \times 10 \times 10 \times 10)}$ or $\frac{1}{10\,000}$ or 0.0001

Note: A number, symbol or unit with an index can be moved from a numerator to a denominator or vice versa by changing the sign of the index.

$\frac{1}{10^{-2}} = 10^2$

$\frac{1}{m^{1/2}} = m^{-1/2}$

$s^{-2} = \frac{1}{s^2}$

Note: Although the laws you will meet at advanced level generally involve indices that are simple whole numbers or simple fractions, indices can be any value. Don't be surprised if the results of a practical investigation suggest that two quantities are related by an index with a number of decimal places.

Note: What if the index is 0?

Anything to the power 0 = 1, so

$10^0 = 1, \quad 2^0 = 1, \quad x^0 = 1$

Questions to try

1. Write down the following without using indices:

   (a) 10^6 (b) 10^{-3} (c) $1/10^2$ (d) 12^2

   (e) $16^{1/2}$ (f) $1/(25^{1/2})$ (g) 3^0 (h) d^2

2. Write the following using indices:

   (a) 10 000 (b) 0.0001 (c) 1/1000 (d) 1/0.01

   (e) $10 \times 10 \times 10 \times 10 \times 10 \times 10$

   (f) $1/10 \times 1/10 \times 1/10$

   (g) $0.1 \times 0.1 \times 0.1 \times 0.1$

   (h) $0.1 \times 0.001 \times 0.01 \times 10$

If you want to know more about:
Reciprocals see page 34

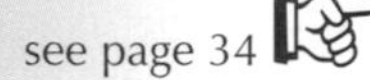

Although you will do most of your work using calculators, you should be able to do a rough check of your processing. The rules for indices apply whether the indices are attached to numbers, symbols for physical quantities or units for quantities.

What if you have something like $5^2–3^2$?
In this case the bases are different so first you have to work out each part separately and then do the subtraction:

$5^2 - 3^2 = 25 - 9 = 16$

$4^2 + 3^3 = 16 + 27 = 43$

What happens when you have something like $(10^5)^{1/2}$?
In this case multiplying the indices produces a fractional index $10^{5/2}$ (or $10^{2.5}$). This is correct but not very useful because for most practical purposes the index should be a whole number. This is covered more fully on page 11.

Multiplying with indices

If numbers that are in the **same base** are to be multiplied you can **add** the indices:

$10^6 \times 10^4 = 10^{(6+4)} = 10^{10}$

$m^2 \times m = m^{(2+1)} = m^3$

What if you have something like $(10^2)^3$?
This is the same as $10^2 \times 10^2 \times 10^2$ which is 10^6. The rule to use in this case is therefore to multiply the indices.

Note: Units are dealt with in the same way as numbers (see page 12). If you have a number of terms that are all squared, remember to deal with each term.

$(\text{mol dm}^{-3})^2 = \text{mol}^2\ (\text{dm}^{-3})^2 = \text{mol}^2\ \text{dm}^{-6}$

What does something like $(10^6)^{1/2}$ or $(10^9)^{1/3}$ mean?
The $\frac{1}{2}$ (or 0.5) means that you need to find the square root of the term, in this case the bracketed number.

As above, you multiply the two indices (which in this case is the same as halving the index). So

$10^{6 \times 1/2} = 10^3$

In $(10^9)^{1/3}$, the $\frac{1}{3}$ means the cube root of the bracketed term is to be found:

$(10^9)^{1/3} = 10^{9 \times 1/3} = 10^3$

So the cube root of 10^9 is 10^3.

Dividing with indices

When dividing, the indices are **subtracted**. Note that the index of the number below the dividing line (the denominator) is subtracted from the number above the line (the numerator).

$\frac{10^6}{10^4} = 10^{(6-4)} = 10^2 = 100$

$\frac{10^4}{10^6} = 10^{(4-6)} = 10^{-2} = 0.01$

$\frac{10^{-19} \times 10^{-19}}{10^{-12} \times 10^{-10} \times 10^{-10}} = \frac{10^{-38}}{10^{-32}}$

$= 10^{-38-(-32)} = 10^{-38+32} = 10^{-6}$

Using indices with units

Understanding indices is important when dealing with units. In the SI system, the correct way to express units is to use indices and not the 'slash' (/) so m/s should be written as ms^{-1}.

The unit m^2 means $(\text{metre})^2$ or (metres × metres) so a length has been multiplied by a length to obtain the result, which is an area.

The unit m s^{-2} means $\frac{\text{metres}}{\text{seconds}^2}$ or $\frac{\text{metres}}{\text{seconds} \times \text{seconds}}$

We would say this unit as 'metres per second squared' or 'metres per second per second'. This is the unit of acceleration.

Questions to try

1. Simplify the following:

(a) $5^2 \times 5^2$ (b) $6^2 - 4^2$ (c) $10^3 \times 10^6$

(d) $\frac{10^{23}}{10^5}$ (e) $\frac{10^{-19}}{10^{-27}}$ (f) $\frac{\text{m}^3}{\text{m}^2}$

(g) $\frac{r^3}{r^3}$ (h) $(4^2)^{1/2}$ (i) $(9^{1/2})^3$

(j) $(10^{-2})^3$ (k) $(\frac{1}{2})^3$ (l) $(10^3)^{1/3}$

(m) $\frac{1}{(10^8)^{0.5}}$

2. Simplify the following units:

(a) $\frac{\text{Pa}^2}{\text{Pa} \times \text{Pa}^3}$

(b) $\frac{\text{kg m s}^{-2}\text{m s}}{\text{m}^2\text{K}}$

(c) $\frac{(\text{mol dm}^{-3})^2}{(\text{mol dm}^{-3})(\text{mol dm}^{-3})^3}$

If you want to know more about:

Standard form see page 10
Using your calculator see pages 26–27

Standard form (scientific notation)

Scientists have to deal with very large and very small numbers. For example, the speed of light is 300 000 000 m s^{-1}, the internal diameter (bore) of a capillary that carries blood in the body is about 0.000 008 m, and in 1 dm^3 (0.001 m^3) of water there are 0.000 000 1 mol of hydrogen ions.

Many physical quantities have values that are much larger or much smaller than these. Writing them down in this form becomes impracticable and leads to errors in calculations. Standard form is used to make it easier to handle these very large and very small numbers.

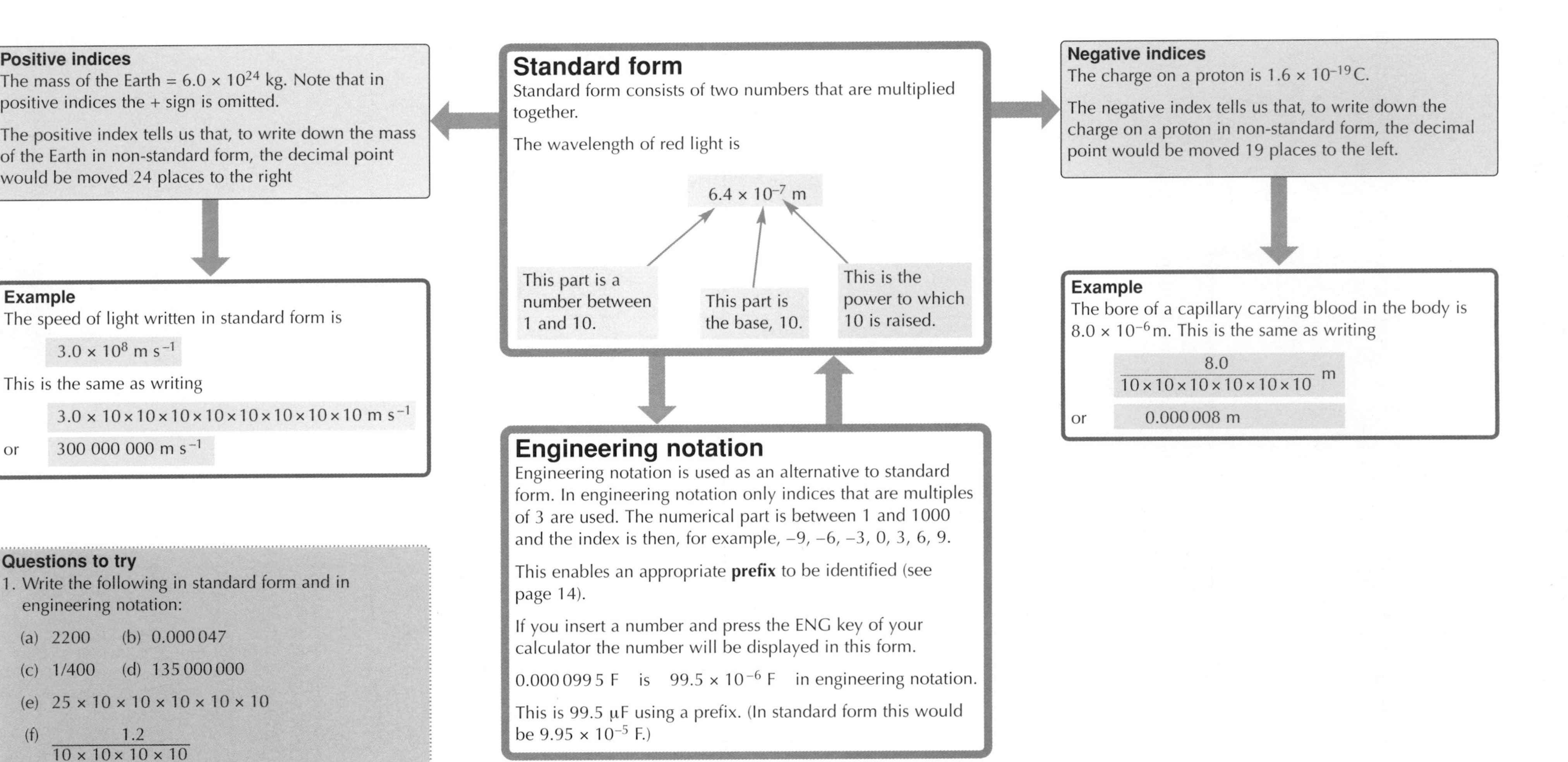

Standard form

Standard form consists of two numbers that are multiplied together.

The wavelength of red light is

6.4×10^{-7} m

This part is a number between 1 and 10.

This part is the base, 10.

This is the power to which 10 is raised.

Positive indices

The mass of the Earth = 6.0×10^{24} kg. Note that in positive indices the + sign is omitted.

The positive index tells us that, to write down the mass of the Earth in non-standard form, the decimal point would be moved 24 places to the right

Example

The speed of light written in standard form is

3.0×10^{8} m s^{-1}

This is the same as writing

$3.0 \times 10 \times 10 \times 10 \times 10 \times 10 \times 10 \times 10 \times 10$ m s^{-1}

or 300 000 000 m s^{-1}

Negative indices

The charge on a proton is 1.6×10^{-19} C.

The negative index tells us that, to write down the charge on a proton in non-standard form, the decimal point would be moved 19 places to the left.

Example

The bore of a capillary carrying blood in the body is 8.0×10^{-6} m. This is the same as writing

$$\frac{8.0}{10 \times 10 \times 10 \times 10 \times 10 \times 10} \text{ m}$$

or 0.000 008 m

Engineering notation

Engineering notation is used as an alternative to standard form. In engineering notation only indices that are multiples of 3 are used. The numerical part is between 1 and 1000 and the index is then, for example, –9, –6, –3, 0, 3, 6, 9.

This enables an appropriate **prefix** to be identified (see page 14).

If you insert a number and press the ENG key of your calculator the number will be displayed in this form.

0.000 099 5 F is 99.5×10^{-6} F in engineering notation.

This is 99.5 μF using a prefix. (In standard form this would be 9.95×10^{-5} F.)

Questions to try

1. Write the following in standard form and in engineering notation:

   (a) 2200 (b) 0.000 047

   (c) 1/400 (d) 135 000 000

   (e) $25 \times 10 \times 10 \times 10 \times 10 \times 10$

   (f) $\dfrac{1.2}{10 \times 10 \times 10 \times 10}$

2. Write the following in non-standard form:

   (a) 3.3×10^{2} (b) 4.8×10^{-6}

   (c) 1.6×10^{8} (d) 2.7×10^{-1}

If you want to know more about:

Prefixes see page 14

Learning the rules that are applied when working with data in standard form will enable you to make a reasonable estimate of the result of substituting data in an equation. Such calculations are useful when presenting scientific arguments in reports. It is also important to check that calculator processing has been carried out correctly.

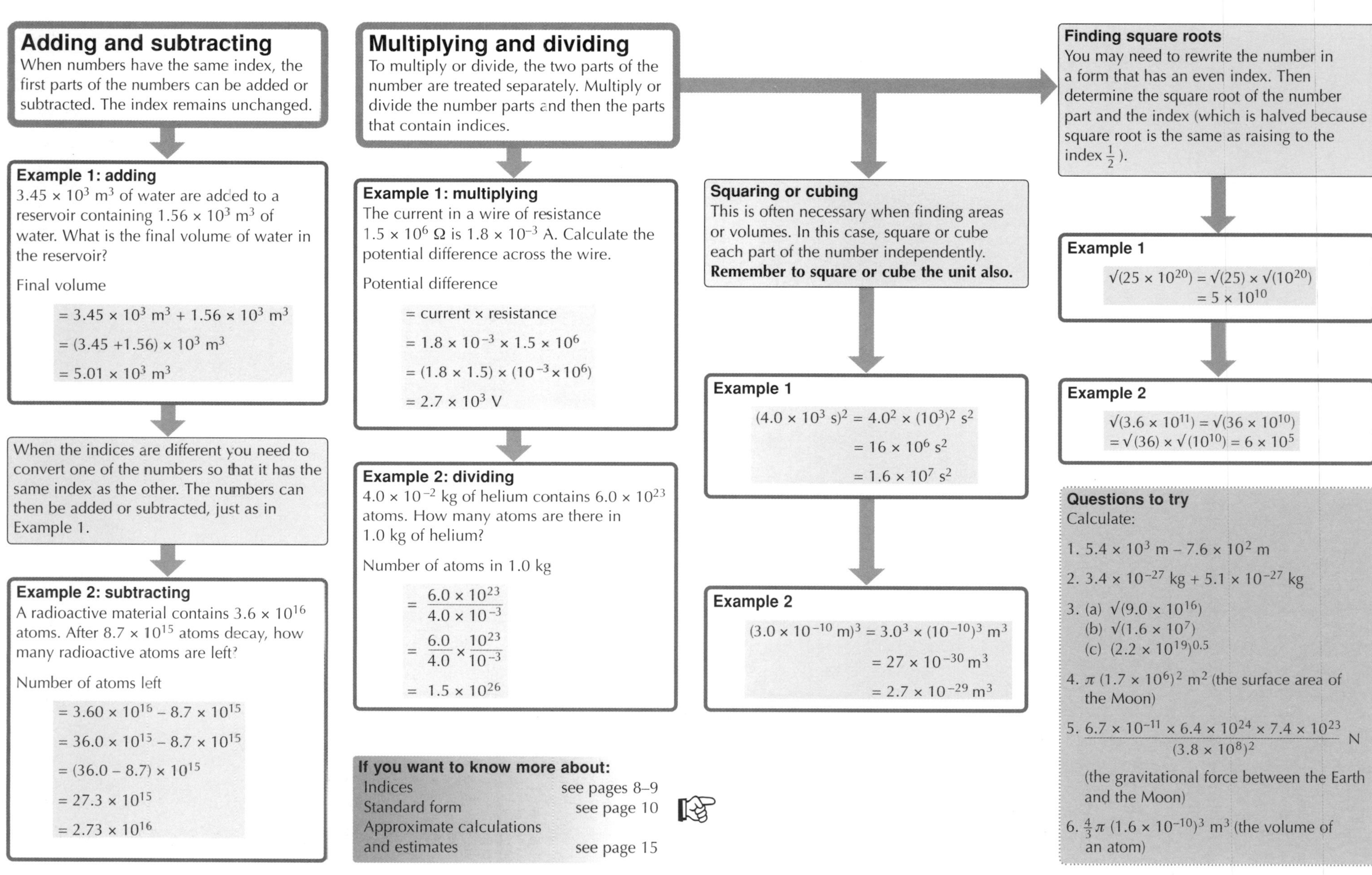

Adding and subtracting

When numbers have the same index, the first parts of the numbers can be added or subtracted. The index remains unchanged.

Example 1: adding

3.45×10^3 m^3 of water are added to a reservoir containing 1.56×10^3 m^3 of water. What is the final volume of water in the reservoir?

Final volume

$= 3.45 \times 10^3\ \text{m}^3 + 1.56 \times 10^3\ \text{m}^3$

$= (3.45 + 1.56) \times 10^3\ \text{m}^3$

$= 5.01 \times 10^3\ \text{m}^3$

When the indices are different you need to convert one of the numbers so that it has the same index as the other. The numbers can then be added or subtracted, just as in Example 1.

Example 2: subtracting

A radioactive material contains 3.6×10^{16} atoms. After 8.7×10^{15} atoms decay, how many radioactive atoms are left?

Number of atoms left

$= 3.60 \times 10^{16} - 8.7 \times 10^{15}$

$= 36.0 \times 10^{15} - 8.7 \times 10^{15}$

$= (36.0 - 8.7) \times 10^{15}$

$= 27.3 \times 10^{15}$

$= 2.73 \times 10^{16}$

Multiplying and dividing

To multiply or divide, the two parts of the number are treated separately. Multiply or divide the number parts and then the parts that contain indices.

Example 1: multiplying

The current in a wire of resistance $1.5 \times 10^6\ \Omega$ is 1.8×10^{-3} A. Calculate the potential difference across the wire.

Potential difference

$= \text{current} \times \text{resistance}$

$= 1.8 \times 10^{-3} \times 1.5 \times 10^6$

$= (1.8 \times 1.5) \times (10^{-3} \times 10^6)$

$= 2.7 \times 10^3$ V

Example 2: dividing

4.0×10^{-2} kg of helium contains 6.0×10^{23} atoms. How many atoms are there in 1.0 kg of helium?

Number of atoms in 1.0 kg

$= \dfrac{6.0 \times 10^{23}}{4.0 \times 10^{-3}}$

$= \dfrac{6.0}{4.0} \times \dfrac{10^{23}}{10^{-3}}$

$= 1.5 \times 10^{26}$

If you want to know more about:

Indices see pages 8–9

Standard form see page 10

Approximate calculations and estimates see page 15

Squaring or cubing

This is often necessary when finding areas or volumes. In this case, square or cube each part of the number independently. **Remember to square or cube the unit also.**

Example 1

$(4.0 \times 10^3\ \text{s})^2 = 4.0^2 \times (10^3)^2\ \text{s}^2$

$= 16 \times 10^6\ \text{s}^2$

$= 1.6 \times 10^7\ \text{s}^2$

Example 2

$(3.0 \times 10^{-10}\ \text{m})^3 = 3.0^3 \times (10^{-10})^3\ \text{m}^3$

$= 27 \times 10^{-30}\ \text{m}^3$

$= 2.7 \times 10^{-29}\ \text{m}^3$

Finding square roots

You may need to rewrite the number in a form that has an even index. Then determine the square root of the number part and the index (which is halved because square root is the same as raising to the index $\frac{1}{2}$).

Example 1

$\sqrt{(25 \times 10^{20})} = \sqrt{(25)} \times \sqrt{(10^{20})}$
$= 5 \times 10^{10}$

Example 2

$\sqrt{(3.6 \times 10^{11})} = \sqrt{(36 \times 10^{10})}$
$= \sqrt{(36)} \times \sqrt{(10^{10})} = 6 \times 10^5$

Questions to try

Calculate:

1. 5.4×10^3 m $- 7.6 \times 10^2$ m
2. 3.4×10^{-27} kg $+ 5.1 \times 10^{-27}$ kg
3. (a) $\sqrt{(9.0 \times 10^{16})}$
   (b) $\sqrt{(1.6 \times 10^7)}$
   (c) $(2.2 \times 10^{19})^{0.5}$
4. $\pi\ (1.7 \times 10^6)^2$ m^2 (the surface area of the Moon)
5. $\dfrac{6.7 \times 10^{-11} \times 6.4 \times 10^{24} \times 7.4 \times 10^{23}}{(3.8 \times 10^8)^2}$ N (the gravitational force between the Earth and the Moon)
6. $\frac{4}{3}\pi\ (1.6 \times 10^{-10})^3$ m^3 (the volume of an atom)

Units

Physical quantities consist of a number and a unit. The units go with the numbers and may be treated as in an algebraic equation to give a correct unit. They should also be transferred with the numerical values in graphical work.

SI units

This is the system of units used in scientific work. SI stands for le Système International. This system defines a number of base units.

Other units may be written in terms of these but many quantities, such as force and power, are so important that they have their own units and these should be used rather than any other unit.

SI base units		
Mass	kg	kilogram
Length	m	metre
Time	s	second
Temperature	K	kelvin
Current	A	ampère
Amount of substance	mol	mole
Luminous intensity (unlikely to be used)	cd	candela

Examples of derived units

Quantity	Unit in terms of base units	SI unit
Frequency	s^{-1}	Hz (hertz)
Charge	A s	C (coulomb)
Momentum	kg m s^{-1} or N s	No other unit
Power	kg m^2 s^{-3}	W (watt)
Force	kg m s^{-2}	N (newton)

Note: Numbers in equations have no units but constants may or may not have units.

Questions to try

1. Write down the following in SI base units:
   (a) V (unit of potential difference)
   (b) Pa (unit of pressure)
   (c) Ω (unit of resistance)
   (d) J (unit of work)
2. Write down the SI unit of the following:
   (a) $1/(\text{period})^2$ (period in seconds)
   (b) mass × $(\text{distance})^2$
   (c) charge per mole
3. Check whether the following equations could be correct:
   (a) Time constant = resistance × capacitance
   $\left(\text{resistance} = \dfrac{\text{voltage}}{\text{current}} \quad \text{and} \quad \text{capacitance} = \dfrac{\text{charge}}{\text{voltage}}\right)$
   (b) Work done = pressure × $(\text{volume})^2$

Working out units

Giving a unit for the final answer is as important as finding the numerical value.

Units can be derived from an equation relating the quantity to other quantities with known units.

It is often useful in calculations to include the units with all quantities, especially in the learning stage. You can then do the necessary mathematical manipulation to determine a unit for the result of the calculation. Alternatively, you can deal with the units separately.

Note that writing 2.5 W is the same as writing 2.5 × W.

Example 1

Calculate the kinetic energy of a mass of 5.0 kg when it is moving at 3.0 m s^{-1}.

$$\begin{aligned}\text{Kinetic energy} &= \tfrac{1}{2} mv^2 \\ &= \tfrac{1}{2} \times 5.0\ \text{kg} \times (3.0\ \text{m s}^{-1})^2 \\ &= \tfrac{1}{2} \times 5.0\ \text{kg} \times 3.0^2\ \text{m}^2\ \text{s}^{-2} \\ &= 23\ \text{kg m}^2\ \text{s}^{-2}\end{aligned}$$

This is a correct unit for kinetic energy, but energy has its own unit, the joule (J), and this should normally be used.

Using this technique, a correct unit for any quantity that has not been met previously can be determined.

Example 2

How much thermal energy is needed to raise the temperature of 3.0 kg of water by 5.0 K? (The specific heat capacity of water is 4.2 × 10^3 J kg K^{-1}.)

$$\begin{aligned}\text{Energy} &= \text{mass} \times \text{specific heat capacity} \times \text{temperature rise} \\ &= 3.0\ \text{kg} \times 4.2 \times 10^3\ \text{J kg}^{-1}\ \text{K}^{-1} \times 5.0\ \text{K} \\ &= 6.3 \times 10^4\ \cancel{\text{kg}}\ \text{J}\ \cancel{\text{kg}^{-1}}\ \cancel{\text{K}^{-1}}\ \cancel{\text{K}} = 6.3 \times 10^4\ \text{J}\end{aligned}$$

Example 3

Could the following be a correct equation?

$$E = mgh + \tfrac{1}{2} Fs^2$$

Energy = mass × acceleration × height + $\tfrac{1}{2}$ force × distance^2

The unit of energy is J which is kg m^2 s^{-2} (from example 1) and the unit of mgh is (kg)(m s^{-2})(m) = kg m^2 s^{-2}, so this part of the equation is fine.

The unit of $\tfrac{1}{2} Fs^2$ is (kg m s^{-2})(m^2) = kg m^3 s^{-2}. This is not the same as the unit for energy so the equation cannot be correct.

Note: checking units in equations

If an equation is correct, the units of each term in the equation (on both sides) must be the same.

If you want to know more about:

Standard form see pages 10–11
Converting between units see page 13
Prefixes see page 14

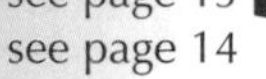

Converting between units

The data obtained in practical work and those given in problems may be presented in a variety of units. In analysis of data it is safer and usually necessary to convert all values to standard form and SI units, without the use of prefixes. This page contains examples of the most frequent conversions required.

Converting areas

A 1 m × 1 m square will contain 100 × 100 squares with a side of length 1 cm.

To convert an area in cm^2 to m^2, multiply by 10^{-4}.

A 1 m × 1 m square will contain $10^3 \times 10^3 = 10^6$ squares of side 1 mm.

To convert an area in mm^2 to m^2, multiply by 10^{-6}.

Converting volumes

By similar reasoning, to convert to m^3:

- Multiply the volume in dm^3 by 10^{-3} (common in chemistry)
- Multiply the volume in cm^3 by 10^{-6}
- Multiply the volume in mm^3 by 10^{-9}

Converting from dm^3 to m^3

The volume of each small cube in the figure is

1 dm × 1 dm × 1 dm
= 1 dm^3

There are 1000 of these in the 1 m^3.

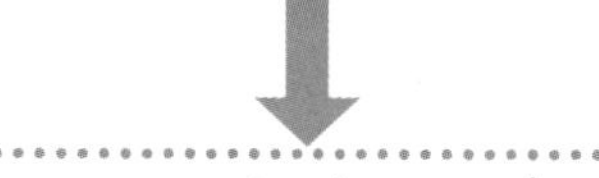

Note: Converting g dm^{-3} to kg m^{-3}

There are 10^3 dm^3 in a m^3, so 1 m^3 contains 10^3 more mass than 1 dm^3.

1 kg = 10^3 g, so the number in kg m^{-3} is the same as the number in g dm^{-3}.

Note:

1 mm^2 means 1 (millimetre)^2 and **not** 1 milli(metre)^2

1 dm^3 means 1 (decimetre)^3 and **not** 1 deci(metre)^3

Converting speeds

Speeds are often given in km h^{-1} (kilometres per hour)

1 km = 10^3 m so 1 km h^{-1} = 10^3 m h^{-1}

There are 60 × 60 s = 3600 s in 1 h, so

10^3 m h^{-1} = 10^3/3600 m s^{-1} = 0.278 m s^{-1}

It is better to remember the process above than to try to remember the conversion factor.

Converting mph to km h^{-1}

Although the km h^{-1} is becoming more familiar in everyday life, many people still find speeds in miles per hour (mph) easier to envisage.

Remember that 1 mph is about 1.6 km h^{-1}. This is worth bearing in mind when thinking about whether an answer is 'about right'.

Note: The reverse conversion from m s^{-1} to km h^{-1} is also useful when considering whether the magnitude of an answer is reasonable.

Speed in km h^{-1} = speed in m s^{-1} × 3.6

Converting pressures

The SI unit of pressure is the pascal (Pa). In many textbooks you will find pressure measured in other units:

1 atmosphere	= 1.01×10^5 Pa
1 bar	= 1×10^5 Pa
760 mm of mercury	= 1.01×10^5 Pa
1 mm of mercury	= 133 Pa

Converting radioactivity decay constants

Decay constants are sometimes given in years^{-1}, min^{-1} or some other time unit. You have to **divide** by the number of seconds in the time unit used.

To convert from year^{-1} to s^{-1} divide by 3.2×10^7 (the number of seconds in a year)

To convert from min^{-1} to s^{-1} divide by 60 etc.

Conversion factors

Conversion required	Multiplying factor	Conversion required	Multiplying factor
g to kg	× 10^{-3}	μF to F	× 10^{-6}
mg to kg	× 10^{-6}	ns to s	× 10^{-9}
μg to kg	× 10^{-9}	kW to W	× 10^3
cm to m	× 10^{-2}	MJ to J	× 10^6
mV to V	× 10^{-3}	GHz to Hz	× 10^9

Questions to try

1. Convert
   (a) 11.1 cm to m (b) 0.035 μs to s
   (c) 35 km to m (d) 110 km h^{-1} to m s^{-1}
   (e) 91.8 MHz to Hz (f) 22.4 dm^3 to m^3
   (g) 2.6 mm^2 to m^2 (h) 1.5×10^3 cm^3 to m^3
   (i) 2.2 min^{-1} to s^{-1} (j) 15 nF to F
2. Convert the following pressures to Pa:
   (a) 2.34 atmospheres (b) 320 mm mercury (mm Hg)
   (c) 1.8 bar (d) 1.0 m Hg (e) 0.013 atmospheres
3. A data book gave the following values in non-SI units. Write down the correct value in SI units.
   (a) 7.6 g cm^{-3} (b) 2.3 N cm^{-2} (c) 24 GJ years^{-1}

Accuracy of data

An advantage of standard form is that the number of significant figures used in the numerical part of the data gives us information about the accuracy of the data. Care must be taken, particularly when using calculators, not to give answers to more significant figures than is reasonable from the data used.

Counting significant figures

To determine the number of significant figures (s.f.), count all the figures written down starting with the first non-zero figure. **You count the last one even when it is zero.**

5.0 is a two s.f. answer (uncertainty 2%).

5.00 is a three s.f. answer (uncertainty 0.2%).

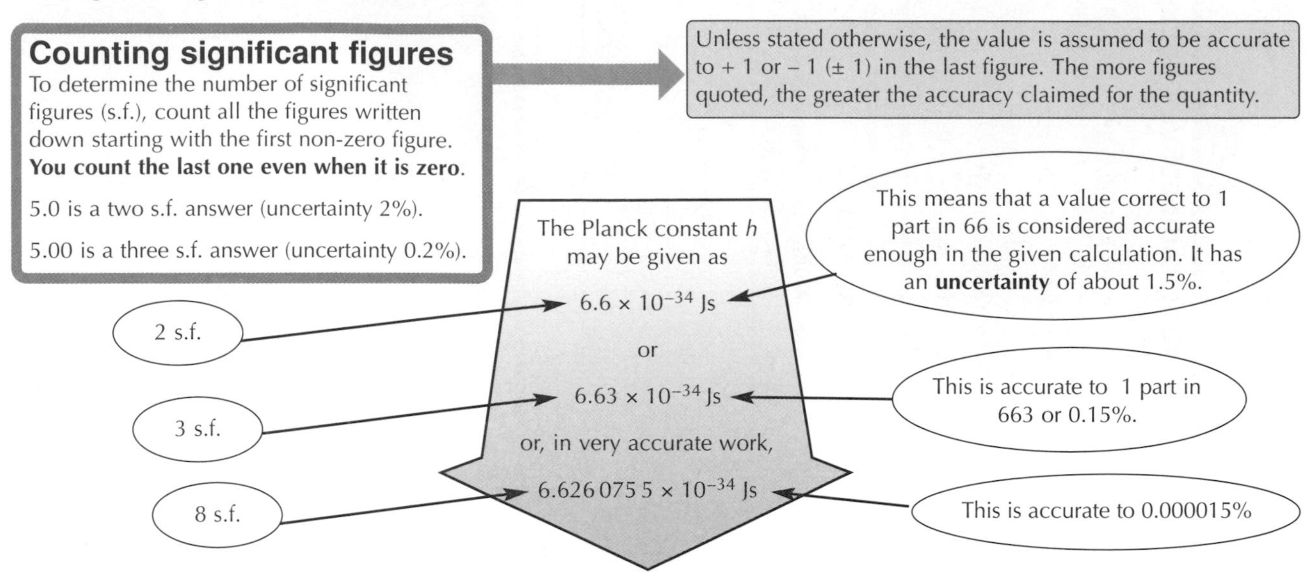

Prefixes

The SI system allows the use of certain prefixes, which signify the index for the base 10. This makes it even easier to write down large and small numbers. You will need to know the meaning of the prefixes.

Note: Although centi and deci are not preferred SI prefixes, they are still in common use in scientific work.

Note: 1 dm^3 is the same as 0.1 × 0.1 × 0.1 or 0.001 m^3. (This is equivalent to 1 litre.)

Note: The prefixes from giga to pico are the important ones for your work, but you are likely to come across the others in your reading.

Note: examples

1.5×10^6 Ω can be written as 1.5 MΩ.

6.3×10^{-6} m can be written as 6.3 μm

1.01×10^5 Pa can be written as 101 kPa or 0.101 MPa.

Symbol	Prefix	Multiplication factor
E	exa	10^{18}
P	peta	10^{15}
T	tera	10^{12}
G	**giga**	$\mathbf{10^{9}}$
M	**mega**	$\mathbf{10^{6}}$
k	**kilo**	$\mathbf{10^{3}}$
d	**deci**	$\mathbf{10^{-1}}$
c	**centi**	$\mathbf{10^{-2}}$
m	**milli**	$\mathbf{10^{-3}}$
μ	**micro**	$\mathbf{10^{-6}}$
n	**nano**	$\mathbf{10^{-9}}$
p	**pico**	$\mathbf{10^{-12}}$
f	femto	10^{-15}
a	atto	10^{-18}

Engineering notation

This uses only powers of 3 and the number part is between 1 and 1000.

In numbers between 1.00×10^3 and 999×10^3 only the number part changes.

The next number in the sequence is 1.00×10^6.

Questions to try

1 Write down the following using prefixes:

(a) 5.0×10^{-9} s (b) 15×10^5 m (c) 2.0×10^{10} N m^{-2} (d) 0.000 063 m^3 (e) 1500 Ω (f) 1.6×10^{-9} s^{-1}

2. How many significant figures are there in each of the following numbers?

(a) 1293 (b) 136.2 (c) 9.7 (d) 0.348 (e) 0.005 847 5 (f) 8.3405×10^3 (g) 3.00×10^{-6}

3. The Faraday constant is $9.648\,530\,9 \times 10^4$ C mol^{-1}. Write it down:

(a) correct to 2 significant figures (b) correct to 3 significant figures (c) in engineering notation

(d) using a suitable prefix. (e) What is the uncertainty expressed as a percentage when quoted as above?

If you want to know more about:

Engineering notation see page 10

Standard form see pages 10–11

Uncertainties see pages 28–29

Approximate calculations (estimates) and order of magnitude

Making an approximate 'order of magnitude' calculation without having to resort to the use of a calculator is a useful skill for a scientist or engineer. An approximate value may be found using measured quantities and rounding them off or by making rough guesses for the magnitude of quantities.

Order of magnitude

The order of magnitude of a quantity is defined by the power of 10 in the number. It gives the approximate size of the quantity.

A change of 1 in the index is a change of one order of magnitude.

Determining the order of magnitude

You need to determine the value of the quantity to the nearest power of 10. The process is as follows:

1. Round off each number in the sum.
2. Calculate the approximate value using the rules on page 11.

Rounding off rules

When the last figure of the number is greater than or equal to 5 it is rounded up; other figures are rounded down:

5.55 to 5.59 become 5.6

5.51 to 5.54 become 5.5

and so

5.57×10^3 becomes 5.6×10^3.

The value could be rounded up to 6. This makes the arithmetic easy enough to do without using a calculator.

For very approximate calculations, it could be rounded to 1×10^4, but take care not to round off too much, especially early on in calculations, otherwise you won't get a very good estimate.

Example 1

A wire resistor has a length of 1.5×10^2 m, and an area of cross section of 4.6×10^{-6} m^2. The resistivity of the metal used is 8.1×10^{-7} Ω m. Determine the approximate value and the order of magnitude of the resistance.

Resistance $= \dfrac{(\text{resistivity} \times \text{length})}{\text{area}}$

$= \dfrac{(8.1 \times 10^{-7}) \times (1.5 \times 10^2)}{4.6 \times 10^{-6}}$

Rounding off leads to

$= \dfrac{(8 \times 10^{-7}) \times (2 \times 10^2)}{5 \times 10^{-6}}$

$= \left(\dfrac{8 \times 2}{5}\right) \times \left(\dfrac{10^{-7} \times 10^2}{10^{-6}}\right)$

$= \dfrac{16}{5} \times 10$

$= 3.2 \times 10\ \Omega$

The approximate value is therefore 30 Ω. (Compare this with the actual value of 26.4 Ω.)

The order of magnitude is 10. Had the answer been 50 or more, the order of magnitude would have been 100. This shows that order of magnitude values are only guides, and approximate values are better estimates.

Note: The order of magnitude of the size of a virus is 10^{-7} m.

The order of magnitude of the distance to the nearest galaxy is 10^{22} m.

The wavelength of X-rays (about 10^{-10} m) is three orders of magnitude smaller than that of visible light (about 10^{-7} m).

Questions to try

Without using a calculator determine an approximate value for each of the following. Write down the order of magnitude in each case.

1. $3500/0.0025$
2. $240^2/85$
3. $\sqrt{30^2 + (95 - 35)^2}$
4. $\dfrac{1}{4\pi^2[(5.1 \times 10^{-3}) \times (2.2 \times 10^{-6})]}$
5. $\dfrac{1.38 \times 10^{-4}}{(3.16 \times 10^4)^2}$
6. $\sqrt{\left(\dfrac{5.2 \times 9.8}{8.1 \times 10^{-2}}\right)}$
7. $\dfrac{0.296 \times 8.31 \times 373}{(1.01 \times 10^5) \times (1.26 \times 10^{-4})}$
8. $\dfrac{1.23}{(1.21 \times 10)^2 \times (1.97 \times 10^{-3})}$
9. $\dfrac{(1.50 \times 10^8)^3 (27.3)^2}{(3.85 \times 10^5)^3 (365)^2}$
10. $\dfrac{(3.2 \times 10^{-19}) \times (1.3 \times 10^{-17})}{4 \times 8.9 \times 10^{-12} \times (1.1 \times 10^{-9})^2}$
11. $\dfrac{4 \times 3.14 \times 150 \times 1.6}{2 \times 1.7 \times 10^{-1}}$

Example 2

Estimate the number of atoms in 1 cm^3 of copper. The density of copper is 8900 kg m^3. The relative atomic mass of copper is 63.5. The Avogadro number is 6.0×10^{23} mol^{-1}.

1 cm^3 = 10^{-6} m^3

Mass of 1 cm^3 = 8900×10^{-6} kg

This is about 9000×10^{-6} kg or 9 g

63.5 g contains 6×10^{23} atoms

9 g contains $\left(\dfrac{9}{63.5}\right) \times 6 \times 10^{23}$ atoms

Rounding off gives

$\left(\dfrac{10}{60}\right) \times 6 \times 10^{23} = 1 \times 10^{23}$

(Actual value = 8.5×10^{22} or 0.85×10^{23})

If you want to know more about:
Indices see pages 8–9
Standard form see pages 10–11

Dealing with negative numbers

Many scientific quantities can be positive or negative. You will learn more about the relevance of the negative sign during studies of individual topics. It is important that you follow the rules for addition, subtraction, division and multiplication of negative numbers to obtain the correct result in a calculation so make sure that you understand them.

Examples of quantities that can have positive or negative values

- **Displacements:** A stone on the ground is assumed to have zero displacement. Upward displacements are positive and downward displacements are negative.

  For a pendulum, displacements to one side of the equilibrium position are positive and those to the other side are negative.
- **Energy:** In chemical reactions, the reacting substances may give energy to the surroundings (an exothermic reaction) or take energy from it (an endothermic reaction). Positive and negative signs distinguish between these processes.

Questions to try

1. An electron in an excited state in a hydrogen atom has energy -5.4×10^{-19} J. It falls to the ground state losing 1.6×10^{-18} J of energy and a photon is emitted as a result. Calculate the energy of the electron in the ground state.
2. The potential energy of a mass of 1.0 kg in a gravitational field is –32 MJ. What is the potential energy of a mass of 1500 kg?
3. Calculate the rise in temperature for a substance that starts at –15.4 °C and ends at 12.8 °C.

A positive number multiplied by a negative number is still negative

The potential energy of an electron that is about 1×10^{-10} m from a proton is

$$\frac{(-1.6 \times 10^{-19}) \times (+1.6 \times 10^{-19})}{4\pi \times 8.9 \times 10^{-12} \times 1 \times 10^{-10}} = -2.3 \times 10^{-18}\ \text{J}$$

When a negative number is squared the result is positive

The current I in a circuit can go in either direction. It may be positive or negative. The power generated is given by I^2R where R is the resistance. Because the current is squared the power is always positive whatever the direction of the current.

Example

In a biology investigation on the length of leaves, one leaf is found to be 9.2 cm long. The mean length of all leaves in the sample is 10.8 cm.

When working out the standard deviation for the experiment, it is necessary to know the *variance*. This is the square of the difference between the observed and the mean value.

Difference = 9.2 – 10.8 = –1.6 cm

Difference^2 = $(-1.6)^2$ = +2.56 cm^2

The square of the difference is a positive number.

'Minus minus' is equivalent to 'plus'

When the temperature of a sample in an experiment falls from 25 °C to –40 °C, the fall in temperature is

$25 - (-40) = 65$ °C

Example

Electron energy levels in an atom are negative.

An electron in an atom is in the energy level at –13.6 eV. It moves up to a level at –1.5 eV.

Energy gained by electron

= –1.5 –(–13.6)

= –1.5 + 13.6

= 12.1 eV

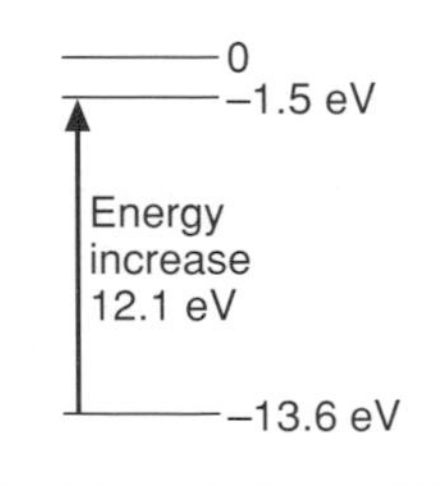

Note: The electron volt (eV) is an energy unit:

$1\ \text{eV} = 1.6 \times 10^{-19}$ J

Taking something from a negative number makes it more negative

It is like taking money from a bank account that is already overdrawn. The overdraft becomes larger!

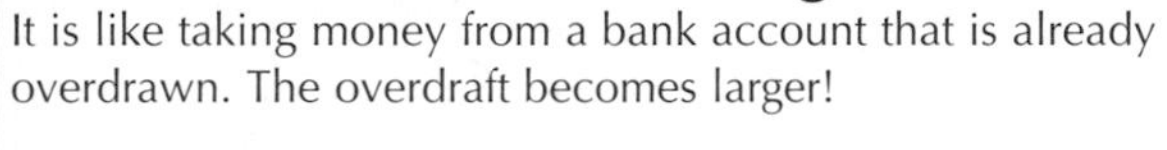

Example

The potential energy of a spacecraft in the Earth's gravitational field is –15 GJ. It loses 12 GJ of potential energy as it falls towards the Earth. What is its new potential energy?

Earth

Diagram not to scale

Spacecraft loses 12 GJ

–27 GJ –15 GJ

New energy = –15 –12 = –27 GJ

If you want to know more about:
Standard form see pages 10–11
Standard deviation see page 74

Ratios

A ratio gives the relative values of two quantities. It tells us how much bigger or smaller one quantity is than another.

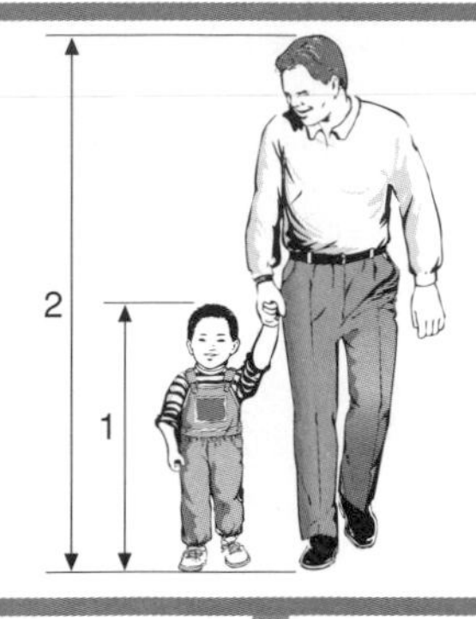

- The ratio of the height of the adult to the height of the child is determined by
  $\dfrac{\text{height of the adult}}{\text{height of the child}}$
- You can say the ratio of the height of the adult to that of the child is 2 to 1. This would be written 2 : 1.
- The 1 is sometimes assumed. We say that the ratio of the height of the adult to that of the child is 2.
- Alternatively you could say $\dfrac{\text{height of the child}}{\text{height of the adult}}$ is 1 to 2, or 1/2, or 0.5 to 1, or simply 0.5.

The ratio has no unit because we are comparing heights.

Example

The mass of Jupiter is 1.9×10^{27} kg. The mass of the Earth is 6.0×10^{24} kg. What is the ratio of the mass of Jupiter to that of the Earth?

$$\text{Ratio of masses} = \frac{1.9 \times 10^{27}}{6.0 \times 10^{24}} = 320$$

Note that the ratio is given to the same number of significant figures as the data.

Questions to try

1. 100 g of lithium contains 7.4 g of the isotope lithium-6 and 92.6 g of lithium-7. Calculate the ratio of lithium-7 to lithium-6.
2. The acceleration of free fall on the Earth is 9.8 m s^{-2}. The ratio of the acceleration of free fall on the Earth to that on the Moon is 6 : 1. What is the acceleration of free fall on the Moon?
3. In a genetics experiment the genotypes are expected to occur in the ratio 9 : 3 : 3 : 1. In this ratio the number with the double recessive genotype is 1.
   (a) What fraction of the population has the double recessive genotype?
   (b) In a total population of 350, how many would be expected to have the double recessive genotype?

When the ratio of two quantities and the value of one of them are known, the other value can be determined.

Example 1

Tensile strain is the ratio of the extension to original length when a wire is stretched.

A wire has an original length of 1.50 m and is stretched so that the strain is 0.012. Calculate the extension.

$$\frac{\text{extension}}{\text{original length}} = 0.012$$

Extension
= original length × 0.012
= 1.50 × 0.012 = 0.018 m

Example 2

The ratio of the volume of oxygen to the volume of other gases in the atmosphere is 0.26 : 1 The volume of oxygen in one sample is 2.5 dm^3. What is the volume of other gases?

$$\frac{\text{volume of oxygen}}{\text{volume of other gases}} = 0.26$$

$$\frac{\text{volume of oxygen}}{0.26} = \text{volume of other gases}$$

$$\text{Volume of other gases} = \frac{2.5}{0.26} = 9.6\ dm^3$$

Ratios may relate more than two quantities. This occurs where a sample contains a number of different components. The proportion of each quantity in the sample can be determined.

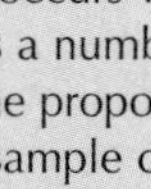

Example

Naturally occurring silicon contains three isotopes. The ratio of the masses of silicon-28 to silicon-29 to silicon-30 is 30 : 1.5 : 1. What is the mass of each isotope in 1.00 kg of silicon?

First find the ratio of the mass of silicon-28 to the total mass. This is

$$\frac{30}{30 + 1.5 + 1} = \frac{30}{32.5} = 0.923$$

So 0.923 of the total mass is silicon-28.

$$\frac{\text{mass of silicon-28}}{\text{total mass of silicon}} = 0.923$$

Since the total mass of silicon is 1.00 kg, the mass of silicon-28 is 0.92 kg.

By similar reasoning the mass of silicon-29 is 0.046 kg and the mass of silicon-30 is 0.031 kg.

If you want to know more about:
Standard form — see pages 10–11
Percentages — see page 18

Percentages

A percentage means 'the number of parts in every hundred'.

Relating an increase or a decrease

Example
Over a year, the number of a species of insect in one area was found to increase by 6%. This means that for every 100 insects counted in one year 106 were counted one year later.

$$\text{Percentage change} = \frac{\text{change}}{\text{original value}} \times 100\%$$

In the first year the number of insects counted was 504 and in the next year the number was 534. The increase was 30.

$$\text{Increase} = \frac{30}{504} \times 100\% = 5.95\% \approx 6\%$$

- The percentage change is 6%.
- The fractional change is 6/100 or 3/50.
- We could say that there were 3/50 more insects.
- The ratio of the increase to the original number = 0.06.

Relating one quantity to the total

Example
A generator is said have an efficiency of 45%. This means that 45 J of every 100 J of fuel used is converted into useful electrical energy.

$$\text{Efficiency} = \frac{\text{useful energy (or power) output}}{\text{energy (or power) input}} \times 100\%$$

The useful output is 2.3 MJ when the energy content of the fuel is 5.1 MJ.

$$\text{Efficiency} = \frac{2.3}{5.1} \times 100\% = 45\%$$

- The percentage efficiency is 45%.
- As a fraction this is 45/100 or 9/20.
- As a ratio it is 0.45 : 1 or 0.45.

Percentages may be converted to fractions or ratios and vice versa

Take care to use the percentage sign when necessary. Otherwise it will be assumed that you mean the ratio.

A percentage can give relative sizes

Example
The abundance of zinc-66 is 50% greater than that of zinc-68. This means that for every 100 parts (kg) of zinc-68 there are 150 parts (kg) of zinc-66. The percentage is relative to zinc-68.

One could also say that the abundance of zinc-68 is 33% less than that of zinc-66 (note that it is not 50% less). For every 150 kg of zinc-66 there are 100 kg of zinc-68. This is a fall of 50 kg. The percentage difference is now relative to zinc-66:

$$\text{Difference} = \frac{50}{150} \times 100\% = 33\%$$

The ratio of zinc-66 to zinc-68 is

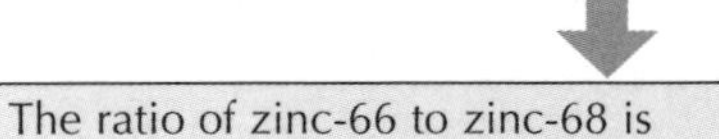

$$\frac{(100\% + 50\%)}{100\%} = 3:2 \text{ or } 1.5:1$$

or, simply, 1.5.

To convert a fraction or ratio expressed as a decimal, multiply by 100

Example
In a biology investigation 15 of 22 students had a pulse rate between 70 and 80 beats per minute (bpm).

- As a fraction this is 15/22 students.
- As a percentage this is 15/22 × 100% = 68%.
- As a ratio, the number with a pulse rate in the range 70–80 bpm to all students tested is 0.68 : 1 or 0.68.
- As a percentage this is 0.68 × 100% = 68%.

Knowing the percentage you can calculate actual values

Example
A compound contains 34.6% copper. How much copper is present in 0.325 kg of the compound?

34.6% means 34.6/100. Therefore the mass of copper is:

$$\frac{34.6}{100} \times 0.325 \text{ kg} = 0.112 \text{ kg}$$

Combining efficiencies

In a power station each stage in the conversion of fuel into electrical energy is inefficient. Typical percentage efficiencies for each stage are shown in the block diagram below.

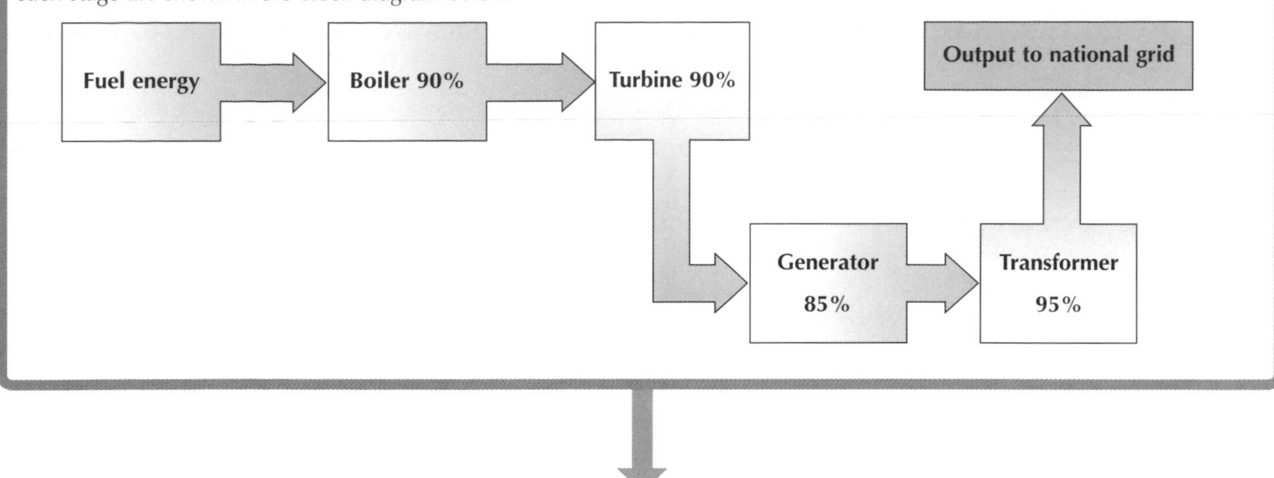

To work out the total efficiency

Each stage further reduces the energy, so to find the overall efficiency multiply together the fraction or decimal efficiencies.

- Write down the efficiency for each stage in fraction or decimal form, e.g. for the first stage 90/100 of the energy input is usefully converted so write the efficiency as 9/10 or 0.90.
- The reduced energy input to the turbine is further reduced so the output is 9/10 × 9/10 or 0.90 × 0.90 = 0.81. Therefore 81% of the fuel energy goes into the generator.
- Repeating this process for each stage the final output of the system is

  9/10 × 9/10 × 85/100 × 95/100

  or

  0.90 × 0.90 × 0.85 × 0.95 = 0.654

  Overall the efficiency of the power station is about 65%.

Determining formulae from percentage composition

Analysis of a compound shows its mass to consist of 40% carbon, 6.7% hydrogen and 53.3% oxygen. This can be used to determine the chemical formula.

- The ratio of the masses combining is the ratio of the percentages 40 : 6.7 : 53.3.
- This means that 40 g of carbon would combine with 6.7 g of hydrogen and 53.3 g of oxygen.
- The number of moles of each element combining is

  $$\frac{\text{mass}}{\text{molar mass of the element}}$$

  The ratio of moles combining is

  $$\frac{40}{12} : \frac{6.7}{1} : \frac{53.3}{16}$$
  $$= 3.3 : 6.7 : 3.3$$

- To simplify, divide each part of the ratio by the smallest number. The ratio of the moles of combining is 1 : 2 : 1 carbon, hydrogen and oxygen.

Note: When this process produces a decimal all numbers have to be multiplied or divided by some factor until they are all simple integers (whole numbers).

- The ratio of the moles combining is the ratio of the atoms combining. So the formula for the compound is CH_2O.

Questions to try

1. The output power of a motor is 255 W. The motor has an efficiency of 28%. Calculate the input power.
2. A plant had a height of 150 mm and there had been an 80% increase in its height in a week. What was the height of the plant at the beginning of the week?
3. The ratio of hydrogen-2 to hydrogen-1 atoms occurring naturally is 15×10^{-5} : 1. What percentage of a sample of hydrogen is hydrogen-2?
4. A compound of carbon, hydrogen and chlorine contains 24.2% carbon, 4.0% hydrogen and 71.8% chlorine. Determine a possible formula for the compound. (The molar masses of carbon, hydrogen and chlorine are 12, 1 and 35.5 respectively.)
5. Three stages of a power station have efficiencies of 60%, 40% and 30%. The output of the second stage is 3.5 MW. Determine
   (a) the power output after the third stage
   (b) the overall efficiency
   (c) the power input at the first stage.

If you want to know more about:
Ratios see page 17

Using formulae: data substitution

Equations are used in science to provide a **mathematical model** of the situation that is being investigated. An equation summarizes the behaviour of a system. When all but one of the quantities in an equation are known, the unknown quantity can be found. This is the situation that you will be faced with most often in advanced-level work.

The meaning of symbols

Symbols in equations may represent:

- variables such as a (acceleration) and F (force)
- constants representing the property of a material such as η (viscosity) and k (thermal conductivity)
- universal constants such as c (speed of light) and G (universal gravitational constant)
- numbers such as π (3.14) and e (2.72)

Some equations with the meaning of their symbols are given on pages 92–93. You should keep a list of all the equations you come across in your course, together with the meaning of the symbols in the equations.

Sometimes the same symbol is used for different quantities

Look carefully at the context in which the equation is used. For example:

- c is used to represent specific heat capacity (as in $E = mc\Delta\theta$) or the speed of light (as in $E = mc^2$).
- E is used for electromotive force (e.m.f.), energy and the Young modulus.

Some symbols are used for quantities and units or quantities and prefixes

Take care not to be misled. For example:

- C is used as the symbol for capacitance.
- C is the unit of charge.

In textbooks, the difference is clear because symbols for quantities are in italic type. You will not be able to do this when writing, and it is easy to make mistakes.

If the quantity you want ...

... is the subject of the formula

You are in luck. Put in the numbers and start calculating.

Example

The temperature change when 3.1×10^{-3} kg of sodium hydroxide dissolves in 0.101 kg water is 8.2 K. Determine the enthalpy change when 1 mol (0.040 kg) of sodium hydroxide dissolves in water. (The specific heat capacity of water is 4.2×10^3 J kg^{-1} K^{-1}.)

enthalpy change $= mc\Delta\theta = 0.101 \times 4.2 \times 10^3 \times 8.2$
$= 3.5$ kJ

Enthalpy change per mol $= \dfrac{3.5 \text{ kJ} \times 0.040 \text{ kg}}{3.1 \times 10^{-3} \text{ kg}}$
$= 45$ kJ mol^{-1}

... is not the subject of the formula

You have two options:

1. Substitute the numbers for the quantities you know in the equation and manipulate the numbers.
2. Rearrange the formula so that the quantity you want to calculate is the subject and then substitute the known quantities (see pages 24–25).

Hint: Data for a problem are often given within a paragraph, in a list of data, on a diagram or in a table or graph.
When the data are in a paragraph of writing it is worthwhile jotting down the data with the usual symbols as you read. This may help you to identify the correct equation to use.

If you want to know more about:

Simultaneous equations see pages 22–23
Changing the subject of formulae see pages 24–25

Using the equations of motion

The equations for uniformly accelerated motion are:

$v = u + at$
$s = ut + \frac{1}{2}at^2$
$s = \frac{1}{2}(u + v)t$
$v^2 = u^2 + 2as$

s displacement or distance travelled
u initial velocity, v final velocity
a acceleration, t time taken

These are often referred to as the *suvat* equations.

When solving a problem:

- Write down the list and place the values you know beside the appropriate letter.
- Place a ? beside the unknown quantities.
- Select the equation from the list that contains only one of the unknown quantities.

Example

A truck travelling at 15 m s^{-1} accelerates at 2.1 m s^{-2} for 3.2 s. How far will it travel while accelerating and what is its final speed?

$s = ?$ $v = ?$ $t = 3.2$ s
$u = 15$ m s^{-1} $a = 2.1$ m s^{-2}

$s = ut + \frac{1}{2}at^2$ will leave only s unknown.
$s = 15 \times 3.2 + \frac{1}{2} \times 2.1 \times 3.2^2$ $s = 48.0 + 10.8 = 58.8$ m
Distance travelled is 59 m to 2 s.f.

$v = u + at$ has only v as unknown
$v = 15 + 2.1 \times 3.2 = 21.7$ m s^{-1}
Final speed = 22 m s^{-1} to 2 s.f.

Cancelling in equations

Cancelling with numbers

When using numbers, cancelling is an alternative to working out the value of each term separately and then dealing with other arithmetical processes in turn.

Example 1

$$\frac{1.5 \times 3.4}{4.8 \times 1.5} = \frac{3.4}{4.8} = 0.71$$

In this example 1.5 appears in each term on the top and bottom so it can be cancelled.

Example 2

This equation could occur when using moments.

$$d\,\frac{(1.7 \times \cancel{9.8} \times 3.0) + (2.3 \times \cancel{9.8} \times 1.5)}{1.5 \times \cancel{9.8}} = \frac{(1.7 \times 3.0) + (2.3 \times 1.5)}{1.5}$$

The 9.8 appears in all three terms and can be cancelled. 1.5 appears in only one of the two terms in the numerator so it cannot be cancelled. However, looking for numbers that are multiples of other numbers can help you to simplify equations further. In this case 3.0 = 1.5 × 2 so

$$d = \frac{(1.7 \times \cancel{1.5} \times 2) + (2.3 \times \cancel{1.5})}{\cancel{1.5}} = \frac{1.7 \times 2 + 2.3}{1}$$

$$= 3.4 + 2.3 = 5.7$$

Any power of 10 that is common can be cancelled if it is common to all terms, e.g. if 10^3 appears in all terms it can be cancelled.

Note: a **term** is a group of quantities that are multiplied together.

VI is a term in an equation

$IR + Ir$ are two terms that are added

Cancelling with symbols

A symbol for a quantity can be cancelled in a numerator and a denominator **when it appears in each term**.

Example 1

$P = V^2/R$ and $V = IR$, so

$$P = \frac{IR \times I\cancel{R}}{\cancel{R}} = IR \times I = I^2R$$

R appears twice on the top (in the numerator) and once on the bottom (in the denominator). One R on the top can be cancelled with the one on the bottom.

Example 2

An equation for kinetic energy conservation in an inelastic collision between a particle moving with velocity u and a stationary particle with the same mass is

$$\cancel{\tfrac{1}{2}}\cancel{m}u^2 = \cancel{\tfrac{1}{2}}\cancel{m}v^2 + \cancel{\tfrac{1}{2}}\cancel{m}V^2$$

where v and V are the final velocities of the particles.

$\frac{1}{2}$ and m both appear in each term on both sides of the equation and can therefore be cancelled. The equation becomes $u^2 = v^2 + V^2$.

Cancelling with units

The cancelling rule also applies when you are simplifying units. Treat each unit in the same way as a symbol in an equation.

Example

Simplify the unit in the following equation.

$$\frac{\text{mol dm}^{-3}\,\text{s}^{-1}}{(\text{mol dm}^{-3})^2(\text{mol dm}^{-3})}$$

Working out the lower squared bracket and cancelling we get

$$\frac{\cancel{\text{mol}}\ \cancel{\text{dm}^{-3}}\ \text{s}^{-1}}{\cancel{\text{mol}}\ \cancel{\text{dm}^{-3}}\ \text{mol dm}^{-3}\ \text{mol dm}^{-3}}$$

$$= \frac{\text{s}^{-1}}{\text{mol}^2\,\text{dm}^{-6}} = \text{mol}^{-2}\,\text{dm}^6\,\text{s}^{-1}$$

Questions to try

Simplifying the following

(a) $0.01 = \frac{k(0.1)^2}{2}$

(b) $\frac{GMm}{r^2} = \frac{mv^2}{r}$

(c) $T = \frac{4.0 \times 10^3\omega_1 - 4.0 \times 10^3\omega_2}{2 \times 10^3}$

(d) $B = \frac{4\pi \times 10^{-7} \times 1.5 \times 10^2 \times 2.0 \times 10^{-1}}{2.0 \times 1.5 \times 10^{-1}}$

(e) $\frac{(\text{m s}^{-1})^2}{\text{m}}$

(f) $\frac{\text{kg}}{\text{kg m s}^{-2}\,\text{m}^{-1}}$

(g) $\frac{\text{mol}^{-3}\,\text{dm}^{-3}\,\text{s}^{-1}}{\text{mol dm}^{-3}\,\text{mol dm}^{-3}}$

(h) $\frac{\text{Pa}^2}{\text{PaPa}}$

(i) $\frac{\text{Pa}^2}{\text{Pa(Pa)}^3}$

If you want to know more about:

Indices see pages 8–9
Standard form see pages 10–11
Units see page 12

Simultaneous equations

When using formulae there is usually only one unknown quantity: you make measurements of all the quantities that you can and then the one unknown can be calculated. Sometimes, however, when all things that can be measured have been measured, there may still be two quantities in the equation. You can use simultaneous equations to determine these quantities.

Solving simultaneous equations

The objective is to combine two equations to produce a third equation in which there is only one unknown quantity. There are two ways of achieving this:

1. by substitution
2. elimination of one of the terms by modifying one of the equations.

Because it has only one unknown, the 'third' (derived) equation will allow you calculate the unknown quantity. The other unknown quantity can then be found by substitution in one of the original equations.

Substitution method (examples 1, 2 below)

1. Express one of the unknown quantities in terms of the other, using one of the equations (say A).
2. Substitute this expression in equation B.
3. Solve the equation for the unknown.
4. Substitute this value into A or B to obtain the other unknown quantity.

Note that it does not matter which equation you use in step 1. The right choice, however, can sometimes lead to easier processing. If you run into difficulties with algebra, it is worth going back and trying the other equation.

Note: Data requirements

To determine two unknown quantities you need two sets of measurements.

The two quantities that you are trying to find must stay the same in both experiments.

You will not be expected to deal with more than two unknowns at advanced level.

Elimination method (example 3 below)

The objective here is to modify one or both equations so that one of the terms containing the unknown quantities is the same in both equations. Adding or subtracting each side of both equations gives an equation that contains only one unknown.

1. Multiply one or both equations by a number that results in two identical terms.
2. Add or subtract both sides of the equations. The equal terms will be eliminated.
3. Solve the equation for the unknown.
4. Substitute this value into equation A or B to obtain the other unknown quantity.

Note: A way that works

- Multiply all the terms in equation A by the number associated with the term you are trying to eliminate in equation B.
- Multiply all the terms in equation B by the number associated with the term you are trying to eliminate in equation A.

This will mean that this term has the same numerical part in both equations. Note that this will not necessarily produce the simplest equations, but it will always work.

Note: How do you know whether to add or subtract the equations?

- If the terms that you are trying to eliminate have the same sign, subtract the two equations.
- If the terms have opposite signs, add them.

Working with brackets

When an equation includes bracketed terms you may need to remove the brackets. Alternatively, to simplify an equation, you may need to introduce a bracket. Remember that you have to work out the brackets before doing any other part of the manipulation.

Equations such as $Ft = m(v - u)$?

You have to remember to multiply each term inside the bracket by the term outside the bracket so that the equation becomes $Ft = mv - mu$.

Adding brackets

A factor that is common to one or more terms in an equation can be taken outside a bracket. For example, the equation

$E = IR + Ir$ becomes $E = I(R + r)$

because I is common to both terms.

What about square roots?

Since the terms inside the brackets are multiplied together, you take the square root of each term in turn.

$\sqrt{(LC)} = \sqrt{L} \times \sqrt{C}$ $(LC)^{1/2} = L^{1/2} \times C^{1/2}$

Note that you cannot simplify $\sqrt{(M + m)}$

How do you square when there are added terms in a bracket?

The rule is that you **add together**:

- the square of the first term
- twice the product of the two terms
- the square of the second term.

The need to do this is frequently linked with small changes. When the speed of a body increases from v to $v + \Delta v$, the velocity squared changes to $(v + \Delta v)^2$. This is equal to $(v + \Delta v)(v + \Delta v)$. You have to multiply each term in the second bracket by each term in the first:

$$v(v + \Delta v) + \Delta v(v + \Delta v) = v^2 + v\Delta v + \Delta v v + \Delta v \Delta v$$
$$= v^2 + 2v\Delta v + (\Delta v)^2$$

If $\Delta v << v$ then $(\Delta v)^2$ will be negligible and the **change** in the square of the speed will be

$$v^2 + 2v\Delta v - v^2 = 2v\Delta v$$

Example 1: Bond energy

In the atomization of ethene, one carbon–carbon (CC) bond and four carbon–hydrogen (CH) bonds are broken. The total energy required to break all the bonds is 2000 kJ mol^{-1}. Butane has three CC bonds and 10 CH bonds. The energy needed to atomize butane is 5170 kJ mol^{-1}.

Calculate the energy required to break a CC and a CH bond.

CC and CH are the unknown quantities.

(CC) + 4(CH) = 2000 kJ mol^{-1} Equation A
3(CC) + 10(CH) = 5170 kJ mol^{-1} Equation B

These are simultaneous equations: there is only one set of values for CC and CH for which both equations are true.

1. Using equation A,

   (CC) = 2000 – 4(CH)

2. Substitute this for (CC) in equation B,

   3[(2000 – 4(CH)] + 10(CH) = 5170

3. Solve the equation:
   - Work out the brackets,

     6000 – 12(CH) + 10(CH) = 5170

   - Combine the CH terms,

     6000 – 2(CH) = 5170

   - Rearrange

     –2(CH) = 5170 – 6000
     2(CH) = 830

Energy of CH bond = 415 kJ mol^{-1}.

4. Now substitute the value for CH in equation A,

   CC = 2000 – (4 × 415)

Energy of CC bond = 340 kJ mol^{-1}.

Example 2: Einstein's photoelectric equation

$hf = \phi + KE_{max}$

h is the Planck constant, f is the frequency of the incident radiation, ϕ is the work function and KE_{max} is the maximum kinetic energy of the emitted electrons.

In an experiment using the same surface, radiation of frequency 6.4×10^{14} Hz caused the emission of electrons of energy 1.2×10^{-19} J, and radiation of frequency 6.7×10^{14} Hz caused the emission of electrons of energy 1.4×10^{-19} J. Determine the work function of the surface and a value for the Planck constant.

The two simultaneous equations are

$h \times 6.4 \times 10^{14} = \phi + 1.2 \times 10^{-19}$ A

$h \times 6.7 \times 10^{14} = \phi + 1.4 \times 10^{-19}$ B

From A, $\phi = h \times 6.4 \times 10^{14} - 1.2 \times 10^{-19}$

Substituting in B,

$h \times 6.7 \times 10^{14} = (h \times 6.4 \times 10^{14} - 1.2 \times 10^{-19}) + 1.4 \times 10^{-19}$

Rearranging gives

$h \times 6.7 \times 10^{14} - h \times 6.4 \times 10^{14} = 1.4 \times 10^{-19} - 1.2 \times 10^{-19}$

This gives

$0.3 \times 10^{14}\, h = 0.2 \times 10^{-19}$

$h = \dfrac{0.2 \times 10^{-19}}{0.3 \times 10^{14}} = 6.7 \times 10^{-34} \approx 7 \times 10^{-34}$ J s

Substituting in A

$6.7 \times 10^{-34} \times 6.4 \times 10^{14} = \phi + 1.2 \times 10^{-19}$

$\phi = 3.1 \times 10^{-19}$ J

Example 3

In these circuits each cell has the same e.m.f. E and internal resistance r. The currents are as shown.

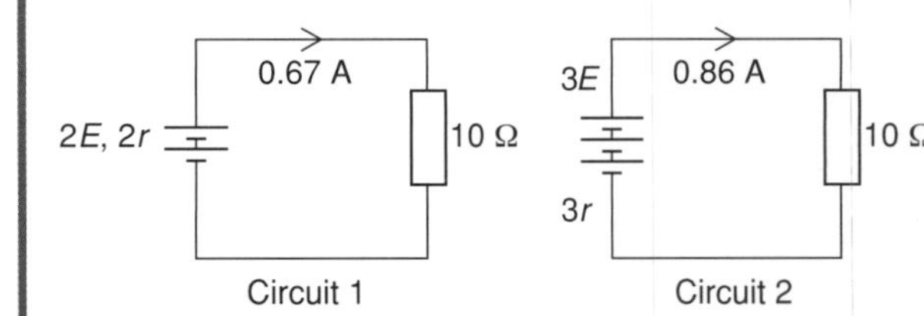

The equations relating E and r for each circuit are:

$2E = 1.34r + 6.7$ (Equation A, for circuit 1)

$3E = 2.58r + 8.6$ (Equation B, for circuit 2)

1. To eliminate the term containing r:

- Multiply all terms in A by 2.58.
- Multiply all terms in B by 1.34.

$5.16E = 3.46r + 17.29$

$4.02E = 3.46r + 11.52$

2. Subtract the second equation from the first.

$(5.16 - 4.02)E = (3.46 - 3.46)r + (17.29 - 11.52)$

$1.14E = 0 + 5.77$

3. Solve the equation:

The e.m.f. of one cell $E = 5.77/1.14 = 5.1$ V.

4. Substitute for E in equation A,

$(2 \times 5.1) = 1.34r + 6.7$

$10.2 - 6.7 = 1.34r$

$r = 2.6\ \Omega$

Note: In example 1 the numbers are small integers. Unfortunately, the numbers are not always as simple as this. Data often involve large and small numbers in standard form, as in example 2 (page 23).

The substitution method is usually easier to handle than the elimination method when numbers involve decimals and powers of 10.

To avoid errors you should take the processing one step at a time and avoid trying to take short cuts.

If you want to know more about:
Standard form see pages 10–11

Changing the subject of formulae

In an equation such as density = mass/volume, density is the subject of the formula. Changing the subject to either mass or volume is called **transposing** the formula. The need to transpose formulae occurs constantly in science. When you can do this effectively there is no need to remember all the variations of a formula, so it is well worth spending time practising this.

Basic principle

For an equation to remain correct, **whatever process is carried out on one side of an equation must be carried out on the other**.

The key to success is not to try to do too many stages at once.

The basic rule applies whether you are working with numbers, symbols or both.

Transposing simple formulae

$$\text{Density} = \frac{\text{mass}}{\text{volume}} \qquad \rho = \frac{m}{V}$$

(a) **Change the subject to *m***

The formula can be written the other way round, so

$$\frac{m}{V} = \rho$$

To make m the subject, multiply m by V. Remember that ρ must also be multiplied by V:

$$\frac{m}{V} \times V = \rho \times V \quad \Rightarrow \quad m = \rho V$$

(b) **Change the subject to *V***

$$\rho V = m$$

Divide both sides by ρ:

$$\frac{\rho V}{\rho} = \frac{m}{\rho} \quad \Rightarrow \quad V = \frac{m}{\rho}$$

Practise with these

The same process can be carried out with formulae such as $V = IR$, $Q = VC$, $\sigma = \frac{F}{A}$, $P = Fv$ and many others. Practise by changing the subject of these formulae.

Do this in stages as in example 1 so that you don't make mistakes.

Note: Don't forget that the units are transposed along with the numerical values.

Working with numbers

Suppose after substituting data you have an equation like this:

$$25\ (\text{J}) = \tfrac{1}{2}\, 3500\ (\text{N m}^{-1}) \times e^2$$

This formula relates energy in a spring to the spring constant and extension e, and you want to determine the value of e.

First work out the numbers on the right-hand side:

$$25\ (\text{J}) = 1750 (\text{N m}^{-1})\ e^2$$

Divide **both sides** by 1750 (N m^{-1}):

$$\frac{25\ (\text{J})}{1750\ (\text{N m}^{-1})} = e^2 \qquad 0.0143\ (\text{J N}^{-1}\ \text{m}) = e^2$$

Take square roots of **both sides** (1 J = 1 N m):

$$0.12\ (\text{N m N}^{-1}\text{m})^{1/2} = e \qquad 0.12\ \text{m} = e$$

What if there are squares or square roots?

When there are squared quantities or a square root is involved, apply the general principles:

- When you take the square root of one side you also take the square root of the other.
- When you square one side you also square the other.

You must remember to square or take the square root of every term in the equation.

Coping with squares

Write the formula $E = \frac{1}{2} CV^2$ so that the subject is V.

First multiply both sides by 2 and divide both sides by C:

$$\frac{2}{C} \times E = \frac{2}{C} \times \tfrac{1}{2} CV^2 \quad \Rightarrow \quad \frac{2E}{C} = V^2$$

Now reverse the equation and take the square root of each side:

$$\sqrt{V^2} = V = \sqrt{\frac{2E}{C}}$$

Coping with square roots

Write the following formula so that the subject is g:

$$T = 2\pi \sqrt{\frac{l}{g}}$$

First square both sides:

$$T^2 = 2^2\, \pi^2 \left(\sqrt{\frac{l}{g}}\right)^2 = 4\pi^2 \frac{l}{g}$$

Multiply both sides by g:

$$T^2 \times g = 4\pi^2 \frac{l}{g} \times g = 4\pi^2 l$$

Divide both sides by T^2:

$$\frac{T^2 g}{T^2} = \frac{4\pi^2 l}{T^2}$$

$$g = \frac{4\pi^2 l}{T^2}$$

Example 1: Einstein's photoelectric equation

$KE_{max} = hf - \phi$

In an experiment to find the Planck constant, h, the following equation arose:

$1.1 \times 10^{-19} = 8.6 \times 10^{14} \times h - 3.9 \times 10^{-19}$

What is the magnitude of h?

First add 3.9×10^{-19} to both sides:

$1.1 \times 10^{-19} + 3.9 \times 10^{-19} = 8.6 \times 10^{14}\, h$

$5.0 \times 10^{-19} = 8.6 \times 10^{14}\, h$

Finally divide both sides by 8.6×10^{14} and reverse the equation:

$$\frac{8.6 \times 10^{14}}{8.6 \times 10^{14}} h = \frac{5.0 \times 10^{-19}}{8.6 \times 10^{14}} = 5.8 \times 10^{-34}$$

so the magnitude of h is 5.8×10^{-34}.

Example 2: Electricity

In electricity the terminal voltage V of a cell is given by

e.m.f. – current × internal resistance: ($V = E - Ir$).

What is r equal to?

First add Ir to each side:

$V + Ir = E - Ir + Ir = E$

Next subtract V from each side:

$V + Ir - V = E - V$
$Ir = E - V$

Finally divide both sides by I:

$$r = \frac{(E-V)}{I}$$

The formula for I is found by dividing by r instead of I in the final stage.

Dealing with angles

In light studies the equation $\sin i / \sin r = \mu$ appears, where μ is the refractive index.

When $\mu = 1.5$ and $i = 55°$, what is the value of r?

$$\frac{\sin 55}{\sin r} = 1.5$$

Invert both sides:

$$\frac{\sin r}{\sin 55} = \frac{1}{1.5} = 0.67$$

Multiply both sides by sin 55:

$\sin r = 0.67 \times \sin 55 = 0.67 \times 0.82 = 0.55$

Find $\sin^{-1}$of both sides:

$\sin^{-1}(\sin r) = r = \sin^{-1}(0.55) = 33°$

Note: the symbol n may also be used for refractive index.

Dealing with exponential equations

In radioactivity, the equation $N = N_0 e^{-\lambda t}$ relates the number of radioactive atoms in a sample to time. Make t the subject of the formula.

First divide both sides by N_0:

$$\left(\frac{N}{N_0}\right) = \left(\frac{N_0}{N_0}\right) e^{-\lambda t} = e^{-\lambda t}$$

Next take the log to the base e (ln) of each side:

$$\ln\left(\frac{N}{N_0}\right) = \ln(e^{-\lambda t})$$

The ln of $e^{-\lambda t}$ is $-\lambda t$, so

$$\ln\left(\frac{N}{N_0}\right) = -\lambda t$$

Finally, dividing both sides by $-\lambda$ and reversing the equation gives:

$$t = \frac{\ln(N/N_0)}{-\lambda}$$

Since $\ln(N/N_0)$ will always be a negative number, the time t will be positive as expected.

Questions to try

1. $v^2 = u^2 + 2as$

   Make the subject of the formula

   (a) v (rather than v^2)

   (b) s

2. $pV = nRT$

   Make the subject of the formula

   (a) p

   (b) n

3. $f = \dfrac{1}{2\pi\sqrt{(LC)}}$

   Make C the subject of the formula.

4. $K_a = \dfrac{[H^+]^2}{[HA]}$

   Make $[H^+]$ the subject of the formula.

5. $3.0 = 6.0 \sin(2\pi f 0.25)$

   (a) Make f the subject of the formula.

   (b) Calculate the value of f.

6. $2.0 = 6.0\, e^{-t/35}$

   (a) Make t the subject of the formula.

   (b) Calculate t.

If you want to know more about:

Standard form see pages 10–11
Exponential changes see pages 47–48
Sines see pages 53, 55
Logarithms see page 65

Using your calculator

Calculator technology, like other electronic technology, is continually improving. Instruments are becoming more and more powerful. For most scientific work a basic scientific calculator is adequate. Calculators differ in the detail of their operation and may have different functions and slightly different labelling. The next two pages include only a few useful tips to assist you in accurate processing of data.

Graphics calculators are being increasingly used and these are useful in practical work where graphs of data can be plotted quickly to examine the general trend before plotting graphs manually. The processing relates to one popular calculator. Whatever calculator you use, it is worthwhile taking time to **read the handbook**. And remember: **practice makes perfect**.

Data processing

Many calculations involve an equation in which data are substituted in a numerator and a denominator:

$$\text{Required answer} = \frac{\text{numerator}}{\text{denominator}}$$

In such calculations it is often safer to work out the numerator and denominator first then do the division:

Work out the numerator first
Press ÷
Include all the denominator in a bracket
Press =

If you try to do the calculation in one go, make sure that you **include brackets**. If you do not do this, things can go wrong. For example:

$$\frac{5+3}{4+6} = \frac{8}{10} = 0.8$$

When you enter the data in the following sequence you do not get the correct answer:

$5 + 3 \div 4 + 6 = 11.75$

This is because the calculator does the division of 3 by 4 and then adds this to 5 and then adds 6 using the BODMAS rules. You should enter

$(5 + 3) \div (4 + 6) =$

You need not include the set of brackets in the numerator in this case, but when there are a number of terms in the numerator it is wise to bracket the whole of the numerator as well as the denominator.

Remember to close every bracket. This may mean having double brackets at the end of long calculations.

Entering numerical data

To enter data in standard form, e.g. 1.5×10^{12}:

Enter	1.5
Press	EXP
Enter	12

For negative exponents (powers of 10) use the special '+/–' key, **not** the arithmetical function (subtraction) key, '–'.

Doing calculations

You must do calculations in the correct order; otherwise, you will not get the correct answer. First you calculate squares, cubes, square roots, etc.

Then work with the data in this sequence:

- **B**rackets
- **O**f
- **D**ivision
- **M**ultiplication
- **A**ddition
- **S**ubtraction.

Remember the initial letters **BODMAS** to remind you of the correct sequence.

A calculator processes data using the same rules.

Use of memory function

An alternative method is to calculate the denominator first and store this in the calculator's memory.

Calculate the denominator
Perform the 'M in' function (Shift M in)
Calculate the numerator
Press ÷
Perform the MR function to recall the denominator
Press =

Log and ln functions

Take care to distinguish between the **log** and **ln** keys on the calculator:

- The log is to base 10. The second function is 10^x.
- The ln is to base e. The second function is e^x.

Angles (radians or degrees?)

Calculators can be programmed to recognize angles entered in degrees or in radians. Mistakes frequently arise because the calculator has been accidentally changed from one to another. Use the **mode** setting key to make sure that it is set correctly for your data. (Deg or rad should appear in the display.) For most purposes it is safer to work in degrees.

Function keys

Keys with two functions:

- The first function is obtained by simply pressing the key.
- The second function (usually printed in a different colour) is obtained by first depressing the INV key and then the key that displays the function.

Note carefully whether you need to use the *first function* or the *second function*. Mistakes can easily be made when calculating sines, cosines and especially tangents. Unless the calculation is for a very small angle, sines and cosines will give an error message, but tangent values can be very large.
For example:

$\tan 25° = 0.466$

$\tan^{-1} 25$ or $\arctan 25 = 87.7°$

Reading and recording the answer

There are two important points to remember.

1. Make sure you quote the answer to an appropriate precision.
2. Take care to include the power of 10 correctly.

Remember that the exponent tells you the power of 10 to include in your answer. A common mistake is to write, for example, 1.23^{-4} when the answer should be 1.23×10^{-4}. The first of these means 0.44 and the second 0.000 123. Quite a difference!

Examples

No attempt has been made in these examples to consider appropriate significant figures. You should look at all data used in any problem and decide what is suitable for the calculation.

1.45 sin 68
- Check that the calculator is in degree mode.
- Enter 68; enter sin; enter ×; enter 1.45; enter =.
- The answer in the display is 0.93.

$3.5^{0.3}$
- Enter 3.5; press x^y (usually a second function); enter 0.3; press =.
- The answer in the display is 1.456.

$\sqrt{2.8 \times 10^{-23}}$
- Enter 2.8; press EXP; enter – using the special +/– key; press √ (usually a second function).
- The answer in the display is 5.29×10^{-12}.

The reciprocal of 1.6×10^{-19}
- Enter 1.6; press EXP; enter – (using the special +/– key); press 19; press $1/x$ (usually a second function key).
- The answer in the display is 6.25×10^{18}.

Log 234
- Enter 234; press log.
- The answer in the display is 2.369.

Antilog 3.592 or $10^{3.592}$
- Enter 3.592; press 10^x (this is usually the second function of the log key).
- The answer in the display is 3908.

$e^{-5.2}$
- Enter 5.2; press the special +/– key to make it negative; press e^x (this is usually the second function of the ln key).
- The answer in the display is 5.517 with –C3 in the exponent part of the display.
- The answer is 5.517×10^{-3}.

Note: Order of pressing keys
On ordinary scientific calculators the number x is entered first and then the function you need (sin, cos, log, ln, $\sin^{-1}$, $\frac{1}{x}$, x^2, etc.).

Note: Doing too much in one step makes it difficult to check calculations. Incorrect answers can arise from trying to do too much at once with the calculator. It is often best to **take calculations one step at a time and to write down intermediate stages**, especially in complex calculations.

Note: In a long calculation it is easy to enter incorrect data. Misplacing a decimal point or entering an incorrect exponent is a common error. It is also surprisingly easy to press × instead of ÷ or vice versa.
Useful checks:
- In short calculations do a quick **order of magnitude calculation** to check that you have not made a slip in entering data.
- In longer calculations do the processing in stages, doing order of magnitude checks at each stage.
- As another check on your processing, it is also wise to **do a calculation twice**, preferably using a different processing sequence.

Note: The number of significant figures for the answer should be consistent with the precision of the data. The calculator will provide answers to as many significant figures as it is programmed to. Some calculators enable you to limit the number of significant figures to, say, 3 or 4. The rule is to give data to the same number of significant figures as the data obtained in an experiment or given in a question.

Questions to try

If you do not obtain the correct answer, try again. If you continue to obtain an incorrect answer, seek advice to find out what you are doing wrong.

Use your calculator to determine:

1. $\frac{45 - 12 \div 2.0}{39 + 24} \times 3.0$
2. $\frac{27.3^2 - 24.8^2}{\sqrt{38}}$
3. $\frac{3.4 \times 10^{-3}}{235} \times 6.0 \times 10^{23}$
4. $\frac{6.6 \times 10^{-34} \times 3.0 \times 10^{8}}{440 \times 10^{-9}} - \frac{1}{2} \times 9.1 \times 10^{-31} \times (4.3 \times 10^5)^2$
5. The reciprocal of 2.34×10^{-5}
6. 1.45^3
7. (a) sin 15.0
   (b) $\cos^{-1} 0.456$
   (c) $\sin(\tan^{-1} 3.8)$
8. $(1.4 \times 10^3)^{3/2}$
9. ln 0.50
10. $\log\left(\frac{2.3 \times 10^3}{4.6 \times 10^{-5}}\right)$

If you want to know more about:
Reciprocals see page 34
Angles see pages 53, 55, 62
Logarithms see pages 65–67

Uncertainty

Measurements made in scientific experiments are never absolutely accurate. Any measured value has an **uncertainty**. The lower the uncertainty the more accurate the measurement is. When measurements are combined to determine another quantity, the errors in each value contribute to the uncertainty in the final result. An important part of experimental work is assessing this uncertainty.

Sources of uncertainty in an experiment

- The design of the instruments
- The calibration of the instruments
- The ability of the user to read the instruments accurately
- The design of the experiment (setting up and procedure)

Care must be taken to minimize each source of error.

Estimating the uncertainty in a reading

The smallest scale division on an instrument is a good guide. Assuming correct calibration, a reading can be made to ± 1 scale division. For example:

- Using a metre ruler a length can be measured to an accuracy of ± 1 mm.
- Using a micrometer a length can be measured to ± 0.01 mm.

EXPRESSING UNCERTAINTIES

Percentage uncertainty

Uncertainties may be quoted as percentages rather than absolute values.

- An uncertainty of 124 ± 1 means 1 in 124.
- Percentage uncertainty = $\frac{1}{124} \times 100\% \approx 0.8\%$

Absolute uncertainty

A length measured as 124 ± 1 mm means that the length lies somewhere in the range 123 mm to 125 mm.

In this case, 1 mm is the absolute uncertainty in the length.

Note: Use of significant figures.

When a length is written as 124 mm, there is an implied uncertainty in the last figure: the uncertainty is assumed to be ± 1 mm. If you want to express a greater uncertainty, you need to state it.

COMBINING UNCERTAINTIES

When you derive another quantity from a set of measurements the quantities are substituted for symbols in a formula. The uncertainty of the value you obtain using the formula depends on the uncertainties of each measurement you have substituted.

Care must be taken in combining uncertainties. Sometimes absolute values are used and sometimes percentage (or fractional) uncertainties.

Adding and subtracting

Total uncertainty = sum of absolute uncertainties

Example: Subtracting

The height of student A is 1.45 ± 0.01 m and the height of student B is 1.36 ± 0.01 m.

The difference in their heights is

$$(1.45 \pm 0.01) - (1.36 \pm 0.01) = 0.09 \pm 0.02 \text{ m}$$

Note that the percentage uncertainty in the difference (≈ 20%) is much greater than the percentage uncertainty in each value (≈ 0.7%).

If the true height of A was at the high end of the range and the height of B at the low end, the difference could have been

$$1.46 - 1.35 = 0.11 \text{ m}$$

The smallest possible difference is $1.4 - 1.37 = 0.07$ m.

Multiplying and dividing

Total uncertainty = sum of percentage uncertainties

Example: Multiplying and dividing

In an experiment on gases the following data together with their uncertainties were obtained. We want to calculate the expected final pressure and its uncertainty.

p_1 = original pressure = $(1.01 \pm 0.01) \times 10^5$ Pa

V_1 = original volume = 3.22 ± 0.01 dm^3

V_2 = final volume = 2.06 ± 0.01 dm^3

Boyle's law states that, at constant temperature, the final pressure of a gas is given by $p_2 = \frac{p_1 V_1}{V_2}$

Final pressure = $\frac{1.01 \times 10^5 \text{ (Pa)} \times 3.22 \text{ (dm}^3)}{2.06 \text{ (dm}^3)}$
$= 1.58 \times 10^5$ Pa

% uncertainty in $p_1 \approx 1\%$

% uncertainty in $V_1 \approx 0.3\%$

% uncertainty in $V_2 \approx 0.5\%$

Total uncertainty in final pressure $\approx 1 + 0.3 + 0.5 \approx 1.8\%$

Absolute uncertainty = $\frac{1.8}{100} \times 1.58 \times 10^5 \approx 0.03 \times 10^5$ Pa

Final pressure = $(1.58 \pm 0.03) \times 10^5$ Pa

Note: In the examples the uncertainties are estimated quantities, so be careful not to express these to too many significant figures. Percentage uncertainties are usually expressed at best to 0.5% when calculated like this. In more advanced work statistical techniques are used to obtain the uncertainties.

What happens to uncertainties when a quantity is squared or cubed?

When a quantity is squared the percentage uncertainty is doubled. When it is cubed, the percentage uncertainty is trebled. The same goes for higher powers: the uncertainty when a quantity is quadrupled is increased four times and so on.

This means that it is very important to measure quantities that are to be raised to a power carefully – even more carefully than other measurements.

Example

A ball bearing has a diameter of 0.0128 ± 0.0001 m. Calculate the volume and its uncertainty.

$$\text{Volume} = \tfrac{4}{3}\pi r^3 = \tfrac{4}{3}\pi 0.0064^3 = 1.098 \times 10^{-6}\ \text{m}$$

$$\text{Percentage uncertainty in } d = \frac{1}{128} \times 100\% = 0.78\%$$

The percentage uncertainty in r is the same as that in d, 0.78%.

Only r is uncertain in the equation for the volume.

$$\text{Uncertainty in the volume} = 3 \times 0.78 = 2.3\%$$

$$\text{Absolute uncertainty in the volume} \approx \frac{2.3}{100} \times 1.098 \times 10^{-6}\ \text{m}^3$$

The final answer is $(1.098 \pm 0.025) \times 10^{-6}$ m^3.

Note that the uncertainty has the same number of decimal places as the calculated volume.

Given that the original data were to three significant figures and the uncertainties would only be approximate in themselves, this should be expressed as $(1.10 \pm 0.03) \times 10^{-6}$ m^3.

Questions to try

1. A student measured the length of a column of water in a capillary tube using a metre rule. The readings at the ends of the columns were 25 mm and 342 mm. Calculate the length of the column and the uncertainty in the value.
2. In an experiment on the solubility of sodium chloride the following data were obtained:

Mass of an evaporating dish	= 28.4 g
Mass of dish + salt	= 33.7 g
Mass of water evaporated	= 14.5 g

   Calculate the mass of salt that dissolved in 100 g of water and the uncertainty in this value.

3. Resistance is voltage/current and the power dissipated is voltage × current. In an experiment the voltage across a resistor was 5.2 ± 0.2 V and the current was 1.25 ± 0.05 A. Determine the resistance and power and the uncertainties in these values.
4. The period T of a simple pendulum was measured five times and the following values were obtained.

   1.24 s, 1.26 s, 1.21 s, 1. 21 s, 1.24 s

   (a) Determine the mean of the values and estimate the standard error in the mean.

   (b) The length l of the pendulum is 375 ± 2 mm. The acceleration of free fall is $g = 4\pi^2 l/T^2$. Calculate a value for g and determine the percentage and absolute uncertainty in the value.

Why we repeat measurements

Taking further independent measurements and averaging provides a more reliable mean value. The extra readings also reduce the uncertainty in that mean value.

You should note that repeating measurements can minimize only random errors. Systematic errors such as those due to zero errors on measuring instruments should be minimized by other means.

The more uncertainty there is in a single measurement the more benefit there is to be gained by repetition. As a matter of good practice **at least one repetition** should be made of **all** measurements.

The **arithmetic mean** of all the measurements is the best estimate of the size of the quantity.

The uncertainty in this case is the **standard error** (of the mean) and this can be determined by statistical methods. (It is not difficult to do using a calculator. See page 75.)

Estimating the standard error

An estimate of the standard error of a number of readings is adequate for many purposes at advanced level.

First determine the difference between the largest and smallest measurements that were made. Divide this by the number of measurements.

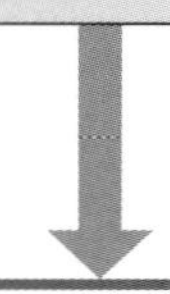

Example

Two students, A and B, measured the period (in seconds) of oscillation of a simple pendulum.

A's measurements: 1.22, 1.26

B's measurements: 1.23, 1.24, 1.22, 1.24, 1.26

What period and uncertainty should each student quote?

A $\text{Mean} = \dfrac{(1.22 + 1.26)}{2} = 1.24$ s

$\text{Uncertainty} = \dfrac{(1.26 - 1.22)}{2} = 0.02$

B $\text{Mean} = \dfrac{(1.23 + 1.24 + 1.22 + 1.24 + 1.26)}{5}$

$= 1.238$

$\text{Uncertainty} = \dfrac{(1.26 - 1.22)}{5} = 0.008$

A could therefore quote 1.24 ± 0.2 s (≈ 1.6%).

B could quote 1.238 ± 0.008 s (≈ 0.7%).

In practice, B would more realistically quote 1.24 ± 0.01 s.

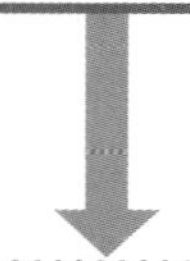

Note: By taking further readings, although student B had the same spread of readings and approximately the same mean, B can be more certain than A of the range in which the true period lies. The repetition has more than halved the uncertainty in the mean value.

Important two-dimensional shapes

The shapes on these pages are the ones you are most likely to encounter. The important features of two-dimensional shapes are the perimeter or circumference and the area of the shape.

PROPERTIES OF TRIANGLES AND QUADRILATERALS

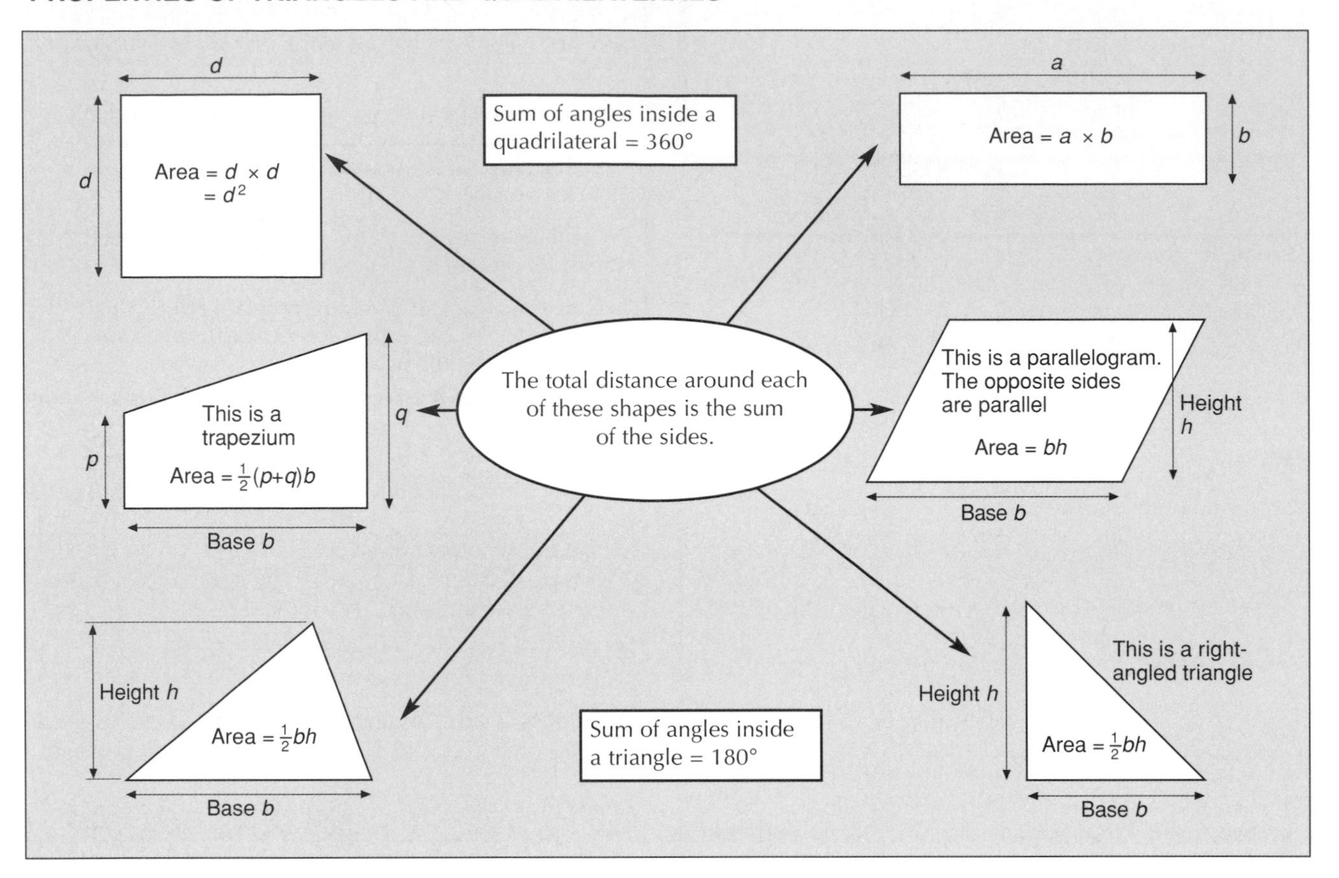

PROPERTIES OF CIRCLES

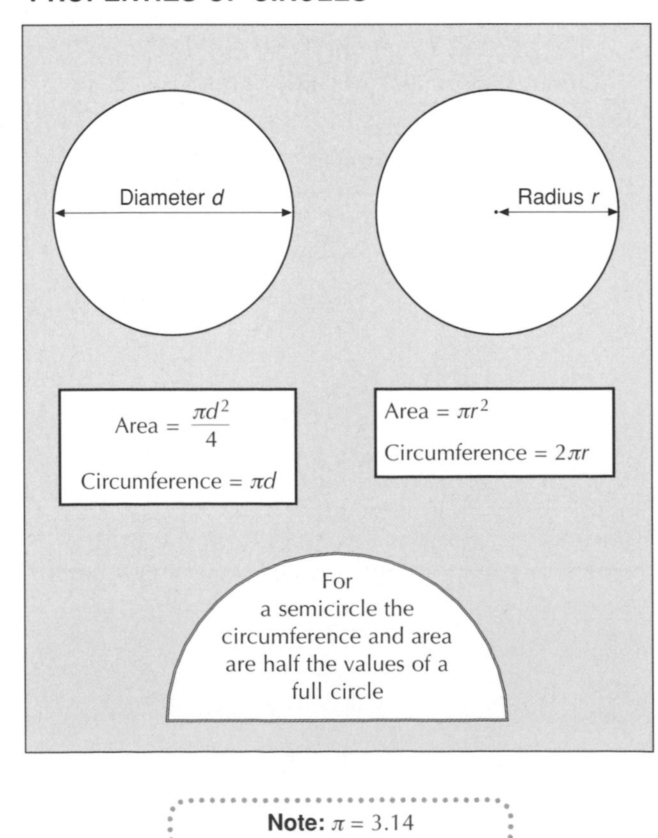

Note: $\pi = 3.14$

Questions to try

1. A rectangle has sides of length 1.4×10^{-2} m and 2.5×10^{-1} m. Calculate the area and the perimeter of the rectangle.
2. A triangle has a base length of 133 mm and a height of 65 mm. Calculate the area in mm^2 and in m^2.
3. The radius of the Sun is 7.0×10^8 m and the diameter of the Earth is 12.8×10^6 m. Calculate the circumference of each of them and the ratio of the circumference of the Sun to that of the Earth.
4. An artery has a diameter of 0.55 mm. Calculate the area through which the blood flows.

If you want to know more about:

Ratios see page 7
Standard form see pages 10–11
Converting between units see page 34

Important three-dimensional shapes

The cube

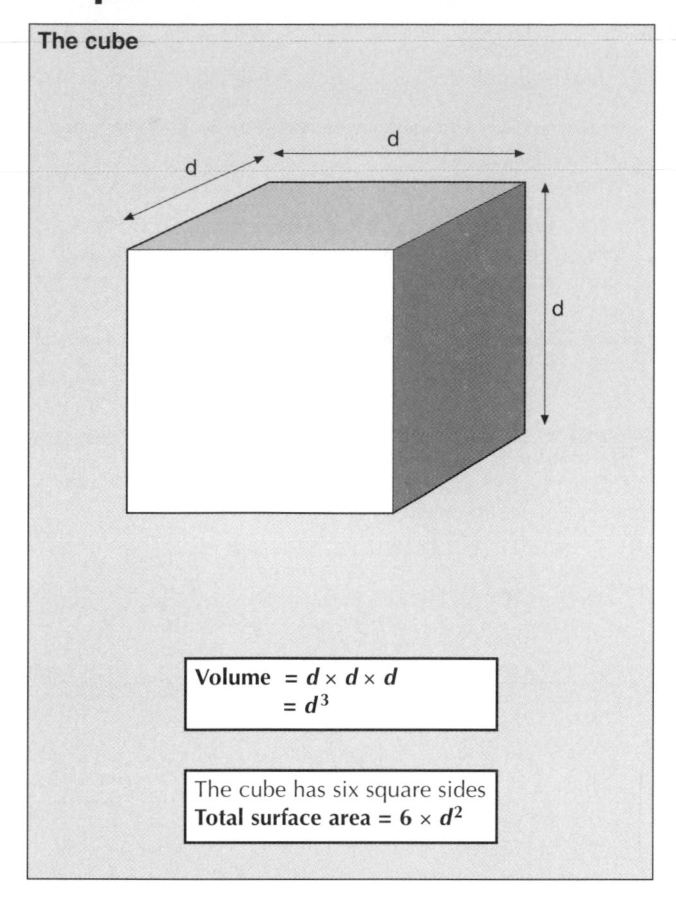

Volume $= d \times d \times d$
$= d^3$

The cube has six square sides
Total surface area $= 6 \times d^2$

Cross-sectional area

- **A rod or bar with square cross-section**

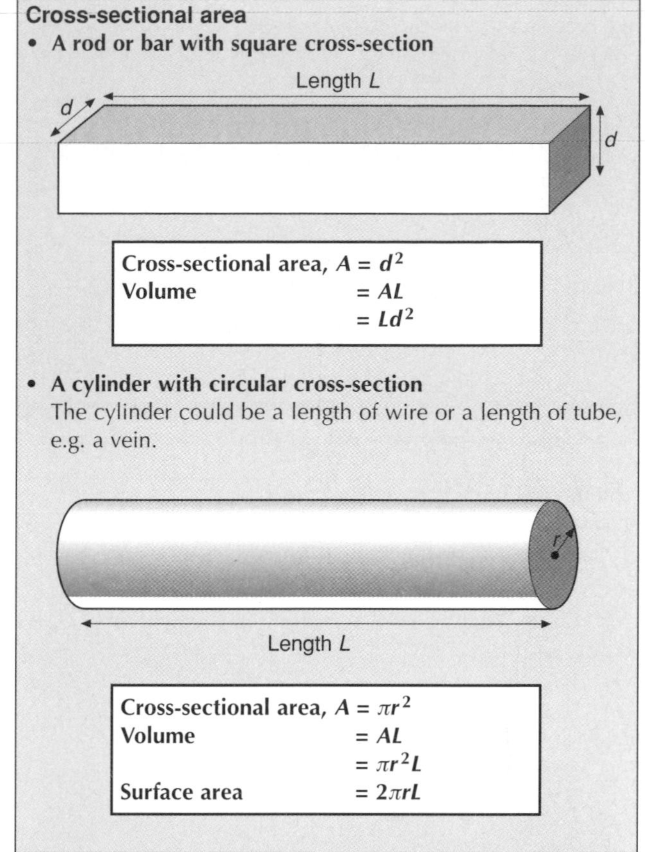

Cross-sectional area, $A = d^2$
Volume $= AL$
$= Ld^2$

- **A cylinder with circular cross-section**
  The cylinder could be a length of wire or a length of tube, e.g. a vein.

Cross-sectional area, $A = \pi r^2$
Volume $= AL$
$= \pi r^2 L$
Surface area $= 2\pi rL$

The sphere

The sphere is a common natural shape, e.g planets and water droplets are approximately spherical and atoms are assumed to be spherical. It is important that you know the equations that relate to the circumference, surface area and volume of a sphere.

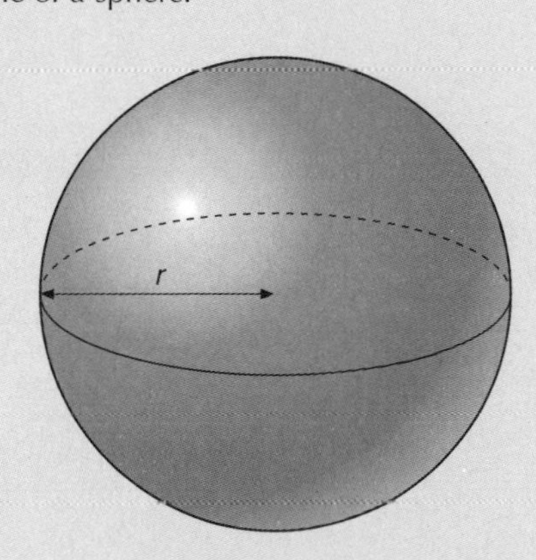

Surface area $= 4\pi r^2$

Volume $= \frac{4}{3}\pi r^3$

Circumference $= 2\pi r$

Questions to try

1. (a) Calculate in m^2 the cross-sectional area of a wire of diameter 0.18 mm.
   (b) Calculate the volume of a 2.5 m length of the wire.
   (c) Calculate the area of the curved surface of the wire.
2. A concrete lintel of length 3.2 m is to be made with a rectangular cross-section 0.15 m by 0.24 m. Calculate the volume of the concrete in the lintel.
3. Calculate the surface area in m^2 and volume in m^3 of a copper cube of side 2.54 cm.
4. Calculate the diameter in m of a spherical flask of volume 1 dm^3.
5. Calculate the surface area in m^2 of a sphere of radius
   (a) 6400 km
   (b) 12 800 km
6. A proton has a radius of about 1.4×10^{-15} m. Calculate the volume of a proton. The mass of a proton is 1.7×10^{-27} kg . Density is mass/volume. What is the density of nuclear matter?
7. A drop of water on a surface is hemispherical and has a diameter of 1.2 mm. Calculate the circumference of the drop in contact with the surface, the area of the surface that is in contact with the drop and the volume of the drop.

If you want to know more about:
Converting between units see page 13

Scaling: making things larger and smaller

Diagrams are often drawn to scale, and scientists and engineers frequently make use of scale models. To represent small objects the drawings or models are made larger than they really are, and those of large objects are made smaller. When using a microscope it is important to know the magnification in order to determine the actual size of the object being investigated.

Magnification using microscopes

Using a microscope produces an image larger than the size of the object. The magnification gives the ratio

$$\frac{\text{length of image}}{\text{length of object}}$$

A magnification of 1500 times means that all lengths are 1/1500 smaller in real life than when viewed in the microscope.

A scale on the graticule of a microscope enables the length and breadth of the image to be determined.

Sketch of an electron micrograph

Photographs of magnified images often show the real dimension as a line on the diagram.

The scale in this figure corresponds to a magnification of about 14 000. The length of the line (14 mm) corresponds to 1 μm.

$$\text{Magnification} = \frac{0.014}{1 \times 10^{-6}} = 14\ 000$$

Over 500 000× magnification is achievable.

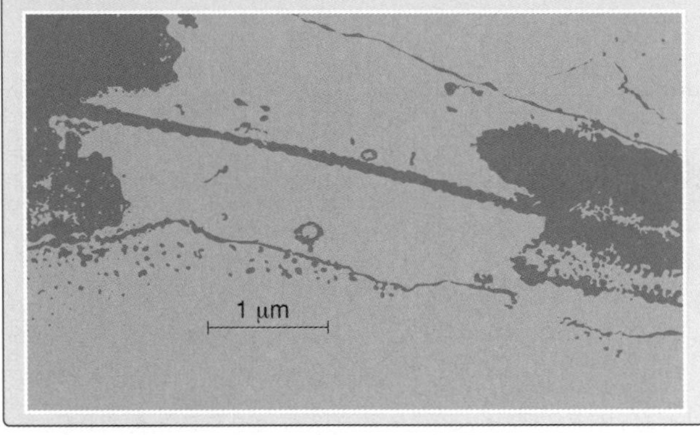

Quoting scales

When scales are used this should be stated clearly.

A scale on a diagram refers to the linear dimensions. This may be given as a ratio as on a map. A scale of 1:100 000 means that each cm on the map represents 100 000 cm or 1000 m or 1 km on the ground.

Representing vector quantities

Scales are used so that vector quantities such as force can be represented on paper.

The scale of this arrow is 1 cm = 5.0 N. This line is 4.3 cm long, so it represents a force of 4.3 × 5.0 = 21.5 N to the right.

Scales on graphs

Graphs are drawn to a scale. The labelling on the axes tells us what 1 cm represents.

EFFECT OF SCALING ON AREA

Area of square	Area of circle	Ratio of linear dimensions	Ratio of areas
1 x 1 = 1	πr^2	1	1
2 x 2 = 4	$\pi(2r)^2 = 4\pi r^2$	2	4
3 x 3 = 9	$\pi(3r)^2 = 9\pi r^2$	3	9

If the linear dimensions are increased *n* times the area increases n^2 times

If all linear dimensions are doubled the area quadruples; if linear dimensions are trebled the area is 9× the original.

EFFECT OF SCALING ON VOLUME

Volume of cube	Volume of sphere	Ratio of linear dimensions	Ratio of volumes
$1 \times 1 \times 1 = 1$	$\frac{4}{3}\pi r^3$	1	1
$2 \times 2 \times 2 = 8$	$\frac{4}{3}\pi(2r)^3 = 8\,(\frac{4}{3}\pi r^3)$	2	8
$3 \times 3 \times 3 = 27$	$\frac{4}{3}\pi(3r)^3$ $27(\frac{4}{3}\pi r^3)$	3	27

If the linear dimensions are increased *n* times the area increases n^3 times

If all linear dimensions are doubled the volume increases 8×; if the linear dimensions are trebled the volume increases 27×.

EFFECT OF LINEAR SCALING ON OTHER PHYSICAL QUANTITIES

When scaling linear dimensions, all physical quantities that are related to length, area or volume will increase (or decrease).

Scaling quantities other than lengths

The effects of doubling, tripling, halving, etc., apply to any quantity. For example, if current I is doubled then I^2 increases four times. If I is reduced to 1/10, I^2 is reduced by a factor of $(1/10)^2$, i.e. to 1/100 of the original value.

Example 1: Mass

A donkey has a similar shape to a dog but its linear dimensions are about 2.5× greater. The mass of the dog is 20 kg. What is the approximate mass of the donkey?

When the linear dimensions increase by 2.5 the volume will increase by $2.5^3 \approx 16\times$. Since

Mass = volume × density

(and the density of a dog and a donkey are about the same), the mass will also increase 16 times, so

Mass of donkey = 16 × 20 kg = 320 kg.

Example 2: Prediction

Scaling can be used to predict the outcome of scaling without knowledge of the actual values.

A formula suggests that

$$X = \frac{AB^2}{C}$$

How will X change when

- A is scaled up by 2
- B is scaled up by 3
- C is scaled down by $\frac{1}{4}$?

$$\text{New value of } X = \frac{(2A) \times (3B)^2}{\frac{1}{4}C} = \frac{2 \times 3^2}{\frac{1}{4}}\left(\frac{AB^2}{C}\right) = 72\left(\frac{AB^2}{C}\right)$$

The new value of X is 72 times larger than the original value.

Example 3: Tensile stress

The tensile stress on a wire is 5.0 kN m^{-2}. What would be the stress for the same force for a wire with linear dimensions a factor of 1.5 larger?

The area of cross-section will be $1.5^2 = 2.25$ times greater.

$$\text{Stress} = \frac{\text{force}}{\text{area}} = \frac{\text{force}}{2.25}$$

The stress changes by a factor of 1/2.25 = 0.444. Therefore

New stress = 0.444 × 5.0 = 2.22 kN m^{-2}

Questions to try

1. Determine the ratio of the volumes of two ball bearings, one of which is 1.5 times the diameter of the other. What will be the ratio of the masses?
2. Saturn has a radius that is 20× that of Pluto. How much larger is the surface area of Saturn than that of Pluto?
3. The period of a pendulum on Earth is 2.0 s. The period of a pendulum is given by $2\pi\sqrt{\frac{l}{g}}$

   What will be the period of a pendulum of half the length on a planet where g is reduced by a factor of 1/6?
4. The pressure between the ends of a pipe is p. The volume flow rate of a fluid of viscosity η in the pipe of length l and radius r is given by the equation

   $$\text{Flow rate} = \frac{\pi p r^4}{8\eta l}$$

   What is the effect on the flow rate of doubling p, halving r and making l three times bigger? (The viscosity remains the same.)

If you want to know more about:
Ratios see page 17
Important shapes see pages 30–31

Reciprocals

The reciprocal of a number = $\dfrac{1}{\text{the number}}$

- The reciprocal of 2 is $\frac{1}{2}$ or 0.5.
- The reciprocal of 0.25 = $\dfrac{1}{0.25}$ = 4.

Numbers with indices

To determine the reciprocal of a number that has an index, simply change the sign of the index.

The reciprocal of 10^6 is $\dfrac{1}{10^6} = 10^{-6}$.

Example 1

What is the period of a wave with a frequency of 90 MHz?

$$\text{Period} = \frac{1}{\text{frequency}}$$

$$= \frac{1}{90 \times 10^6} = \frac{1}{90} \times \frac{1}{10^6}$$

$$= 0.0111 \times 10^{-6} = 1.11 \times 10^{-8}\,\text{s}$$

Example 2

The decay constant of phosphorus-32 is 21 days. Calculate the average lifetime of an atom of phosphorus-32.

Decay constant = 1/average lifetime

$$= \frac{1}{21 \times 24 \times 60 \times 60}\,\text{s}^{-1}$$

$$= \frac{1}{1.81 \times 10^6}$$

$$= \frac{1}{1.81 \times 1/10^6} = 0.55 \times 10^{-6}$$

$$= 5.5 \times 10^{-7}\,\text{s}^{-1}$$

Example 3: Lens formulae

The formula is $\dfrac{1}{f} = \dfrac{1}{u} + \dfrac{1}{v}$

- f is the focal length of the lens.
- u is the object distance from the lens.
- v is the image distance from the lens.

An object is 0.06 m from a lens of focal length 0.08 m. Where is the image?

$$\frac{1}{f} = \frac{1}{u} + \frac{1}{v} \Rightarrow \frac{1}{0.08\text{ m}} = \frac{1}{0.06\text{ m}} + \frac{1}{v}$$

$$\frac{1}{v} = \frac{1}{0.08\text{ m}} - \frac{1}{0.06\text{ m}} = 12.5\text{ m}^{-1} - 16.7\text{ m}^{-1} = -4.2\text{ m}^{-1}$$

Taking reciprocals of both sides gives

$v = -0.23$ m

The image is 0.23 m from the lens. The negative sign in this case means it is on the same side of the lens as the object.

Numbers in standard form

The reciprocal of a number in standard form is the reciprocal of each part multiplied together. The reciprocal of 5.0×10^{-4} is $\dfrac{1}{5.0} \times \dfrac{1}{10^{-4}} = 0.20 \times 10^4 = 2.0 \times 10^3$

Example: Capacitors in series

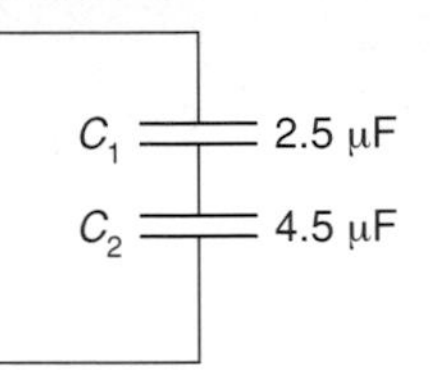

The formula that gives the effective capacitance of capacitors in series is

$$\frac{1}{C} = \frac{1}{C_1} + \frac{1}{C_2} = \frac{1}{2.5 \times 10^{-6}\text{ F}} + \frac{1}{4.5 \times 10^{-6}\text{ F}}$$

Finding the reciprocal of each term on the right-hand side gives

$$\frac{1}{C} = 0.40 \times 10^6\text{ F}^{-1} + 0.22 \times 10^6\text{ F}^{-1} = 0.62 \times 10^6\text{ F}^{-1}$$

Finding the reciprocal of each side gives

$C = 1.6 \times 10^{-6}\text{ F} = 1.6\ \mu\text{F}$

Similar calculations are required for dealing with resistors in parallel (see question 3).

Questions to try

1. Calculate the reciprocal of the following quantities. Include a suitable unit for the reciprocals.

   (a) 1.62 m (b) 220 Ω (c) 2.2×10^{-9} F

   (d) 50 ms (e) 1.52 kHz (f) $2.3 \times 10^5\text{ s}^{-1}$

2. The wave number is 1/wavelength in m. Calculate the wave numbers for waves of the following wavelengths:

   (a) 1500 m (b) 2.9 cm (c) 590 nm

3. The formula for calculating the effective resistance of three parallel resistors is

   $$\frac{1}{R} = \frac{1}{R_1} + \frac{1}{R_2} + \frac{1}{R_3}$$

   Three resistors in parallel have values 7.5 Ω, 5.3 Ω and 3.4 Ω. Calculate the effective resistance of the combination.

4. A lens used in a magnifying glass has a focal length of 0.25 m.

   (a) An object is placed 0.20 m from the lens. Where will the image be formed?

   (b) Where will the image be when the object is 0.45 m from the lens?

If you want to know more about:

Units see pages 12–13

Standard form see pages 10–11

Graphs

A graph is a powerful tool in scientific work. Good graph drawing skills and an understanding of how to use graphs are essential skills for a scientist or engineer.

Using graphs

A graph shows at a glance how one quantity changes with respect to another. It is easy to see fluctuations and trends and, sometimes, to derive an equation relating the quantities.

Dependent and independent variables

- The quantity to be investigated is called the **dependent variable**.
- The factors that you control (either fix or vary) are called the **independent variables**.

The design of the experiment must ensure that only one of the independent variables is changed at a time, the others being kept constant. This is called **controlling the variables**.

Dot to dot or a line of best fit?

In the example below it is not reasonable to draw a line of best fit and the points are simply joined. The temperatures are known only at the times the temperature is taken. What happens at times between the measurements is unknown and cannot be determined from the measurements made.

In other situations, such as in the variation of pressure with volume for a gas (see 'plotting data'), the data are continuous and it is reasonable to draw a line of best fit. The values of pressure and volume between the measured values can be expected to lie on or close to the graph line.

Example

This graph gives a good visual image of how the temperature of a patient varied with time. The trend toward a normal temperature can be clearly seen.

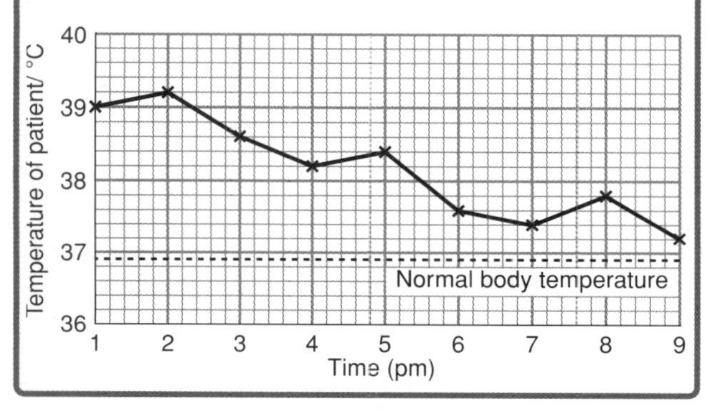

Note: When the graph is a curve like this, drawing a line of best fit may be difficult. It is also hard to see what might happen outside the range. Extrapolation is very unreliable.

In the example here (below) pressure p is the dependent variable. The factors that affect pressure (the independent variables) are the mass of the gas, its temperature and volume. Mass of gas and temperature are kept constant. Volume V is the independent variable that is changed by the experimenter in this case.

Observations of V and the corresponding values of p are made and tabulated.

Construction of the table

A table should include all observations made, including any repeat readings and the average of the readings.

Each column contains a heading with the unit and a power of 10, where appropriate. The numbers in the table are 'pure numbers'. Do not include the unit again.

To obtain the actual value corresponding to the tabulated value, multiply by the appropriate power of 10 and give the unit shown at the head of the column.

Use the same number of significant figures for all the entries in one column and for related columns.

Take care when drawing and reading from graphs to take account of the powers of 10.

First reading of $V/10^{-3}$ m^3	Second reading of $V/10^{-3}$ m^3	First reading of $p/10^5$ Pa	Second reading of $p/10^5$ Pa	Average volume $V/10^{-3}$ m^3	Average pressure $p/10^5$ Pa
3.8*	3.4	0.81	0.83	3.6	0.82**
2.8	2.8	1.03	1.07	2.8	1.05
2.3	2.5	1.22	1.24	2.4	1.23
2.0	2.1	1.41	1.40	2.1	1.41
1.8	1.7	1.64	1.64	1.8	1.64

* This means 3.8×10^{-3} m^3. ** This means 0.82×10^5 Pa.

Plotting data

These results may be plotted directly on graph axes.

- The **independent variable** is usually plotted on the ***x*-axis** or **abscissa** (i.e. the horizontal axis).
- The **dependent variable** is plotted on the ***y*-axis** or **ordinate.**
- This is a graph of p **against** V since: p is on the y-axis; V is on the x-axis.

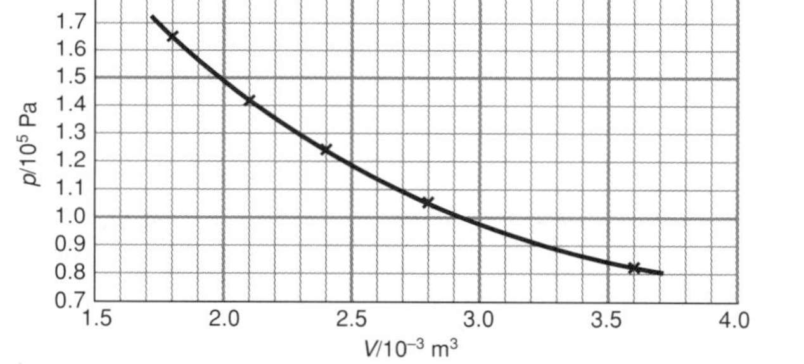

The art of graph drawing

Drawing graphs

- Use a sharp pencil.
- Label both axes with the quantity, the power of 10 where appropriate and the unit. The preferred way is to write, for example, 'Current/10^{-3}A', meaning that each integer number on the axis represents 10^{-3}A.
- Choose a sensible scale: use 1 cm to represent 1 unit, 2 units or 10 units. Note that using 1 cm to represent 3 units makes plotting 0.1 difficult. It may lead to errors when reading from the graph.
- Indicate the scale reading every 1 cm or every 2 cm.
- Plot the points accurately and clearly + or × symbols are best but ⊙ may be used.
- Draw the best line (curve or straight line). The best line is the one that gives as even a distribution of points about the line as possible.

  When data lead to a curve, it can be difficult to draw the best curve through the points. Instruments such as 'flexicurves' and 'universal curves' are useful.
- State on the graph what it represents, i.e. give it a title.

Make full use of the graph paper

Consider using a **false origin** so that points are well spaced. A false origin is one that starts with a value other than 0.

When both scales start at 0, there may be a lot of wasted graph paper which carries no information.

However, when the origin is included, it is possible to check at a glance whether the graph is a straight line through the origin. This is the test for direct proportionality.

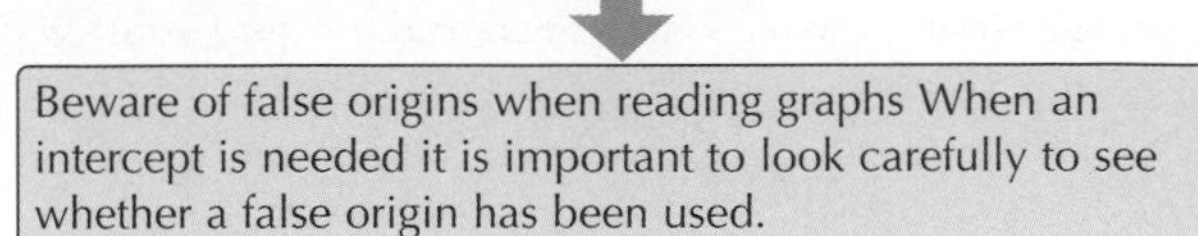
Beware of false origins when reading graphs When an intercept is needed it is important to look carefully to see whether a false origin has been used.

Note: The coordinates of a point are the corresponding x and y values. These are always quoted with the quantity on the x-axis first. The origin has coordinates (0,0).

DRAWING MORE USEFUL GRAPHS

Wherever possible scientists try to find a relationship that gives a straight-line graph

A straight-line graph:

- makes it easier to draw the best line for the data so that the trend is determined more accurately
- improves the accuracy of predicting values other than those measured within the range of the experiment (**interpolation**)
- enables more reliable assessment of what might happen outside the range of the experiment (**extrapolation**).

Possible linear relationships may be found by:

- using knowledge of the laws of physics to predict what the relationship might be (**creating a hypothesis**)
- guessing what the relationship might be by examining the trends in the data and doing rough calculations to predict a law (**empirical analysis**)
- attempting analysis of the data using logarithms.

Using derived quantities

Theory may suggest that **derived values** of a quantity should be plotted to gain the advantages of a straight-line graph.

This means that, if the data in the table include a quantity r, to obtain a straight-line graph it may be necessary to plot a quantity such as $1/r$, $1/r^2$, r^2 or r^3.

Example

Robert Boyle suggested that when the temperature of a fixed mass of gas is constant the pressure is inversely proportional to the volume, $p \propto \frac{1}{V}$.

If this is true, we can predict that a graph of p against $\frac{1}{V}$ should be a straight line through the origin.

$p/10^5$ Pa	$(\frac{1}{V})/10^2$ m^{-3}
0.82	2.8
1.05	3.6
1.23	4.2
1.41	4.8
1.64	5.6

Note: The unit for the reciprocal of volume:

- V is in m^3
- $\frac{1}{V}$ is in m^{-3}

Graph of p against $\frac{1}{V}$ using the above data, calculated from the data in the example on the previous page

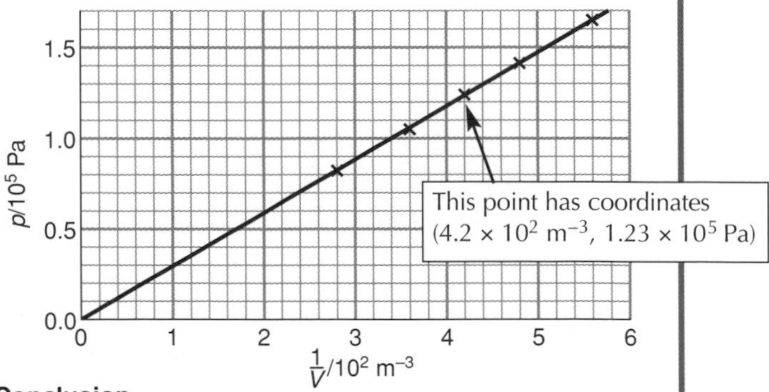

Conclusion

Since the graph produced from the data is a straight line through the origin, the data support the predicted law and Boyle's law is proved to hold for the experiment.

If you want to know more about:

Reciprocals see page 34

Drawing straight-line graphs see pages 42–45

Using graphs

Graphs can provide a wealth of information. Intercepts on axes, gradients and areas between the graph line and the axes may all have physical significance. **Interpolation** and **extrapolation** mean determining the corresponding values of points that were not included in the measurements made in the experiment.

Interpolation

This means finding corresponding values for quantities **within** the range of the experimental data.

These values will have the same uncertainty as the original data and, when the data show a clear trend, they are reliable predictions.

Extrapolation

This means extending the graph at either end in an attempt to predict what happens **outside** the experimental region.

When extrapolating it is assumed that the trend of the data continues unchanged.

You need to be sceptical about conclusions when extrapolating data. The further the line is extended, the less reliable the predictions will be. The factors affecting the relationship between the two quantities may change, so that the trend may change. This is true for both linear and curved relationships.

Intercepts

The intercept is the value of one quantity when the other is zero. For example, you can determine the length of a metal bar when the temperature drops to 0 °C from experimental data that show the trend between 20 °C and 80 °C.

Finding an intercept is an important step in finding an equation relating two quantities.

Measuring the gradient of a straight-line graph

The gradient of a straight-line graph is constant; the rate of change remains the same.

1. Use points **on the line** that are separated by a large distance in order to minimize errors – ideally by at least half the line that has been drawn.
2. Preferably show a large triangle on the graph or make clear in the analysis the coordinates of the points that are used.
3. Always use points that are **on the line**. Do not use tabulated values.
4. Determine the values of change in y (Δy) and change in x (Δx) represented by the sides of the triangle. Remember to use the scales on the axes **not** the length of the lines.
5. Calculate the gradient $\Delta y/\Delta x$.

Remember that the gradient cannot be quoted to more significant figures than the readings taken from the graph for Δy and Δx.

Measuring the gradient of a curved graph

The gradient of a curved graph is changing. In this case it may be necessary to determine the gradient at a point on the curve.

1. Identify the point at which the rate of change is required. If the x-axis is time, this would be a particular time.
2. Draw a **tangent** at that point; that is, a line that just touches the curve. The gradient of this tangent is the rate of change at that point.
3. Make the tangent line as long as is convenient to minimize errors.
4. Proceed as in points 4 and 5 for measuring the gradient of a straight line.

Example 1: Intercept and gradient of a curved graph

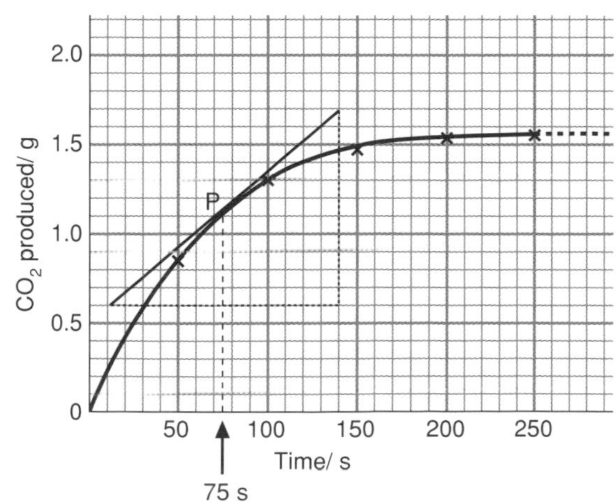

The gradient is changing, so you can only determine the gradient at a particular time:

- The gradient 75s after the start is $\frac{1.10 \text{ g}}{130 \text{ s}} = 0.0085 \text{ gs}^{-1}$.
- The gradient at the origin would give the greatest production rate.

Example 2: Straight-line graph showing the intercept, an interpolated point, an extrapolation and a gradient triangle

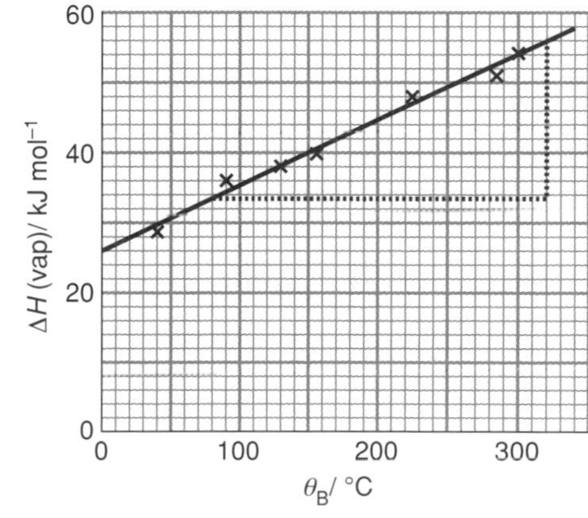

- The intercept is 26 kJ mol^{-1}. This is found by extrapolation.
- A substance with a ΔH(vap) of 44 kJ mol^{-1} would have a boiling point of 190 °C. This is found by interpolation.
- Using the gradient triangle, the gradient is $\frac{22 \text{ kJ mol}^{-1}}{240 \text{ °C}} = 0.092 \text{ kJ mol}^{-1} \text{ °C}^{-1}$.

Rates of change / graphs

Why measure gradients?

A gradient may:

- represent another physical quantity directly
- enable the calculation of a physical constant or another quantity
- lead to an equation that relates the quantities plotted.

Gradient and rate of change

The gradient is the change in the quantity plotted on the *y*-axis divided by the corresponding change in the quantity plotted on the *x*-axis. It tells us the **rate of change** of quantity *y* with quantity *x*. The term **rate** may lead to the misconception that rate of change applies only to changes with respect to time. Although this is often the case, the term rate of change also applies to changes with respect to other independent variables.

On a straight-line graph of the length of a bar plotted against temperature the gradient shows the rate of change of length with temperature. A steeper gradient would show that the rate of change of length with temperature was greater.

To suggest that a steeper gradient meant that the length increased more 'quickly' would not be precise. This would suggest that it did not take a long time to expand. Try to reserve terms like 'quickly' and 'faster' for variations with respect to time to avoid ambiguous statements.

Unit for the gradient

The gradient represents one physical quantity divided by another so it may have a unit. The unit is

$$\frac{\text{the } y \text{ unit}}{\text{the } x \text{ unit}}$$

When the unit of both these quantities is the same the gradient will be a ratio. It will not have a unit.

Example 1: A gradient may represent another physical quantity

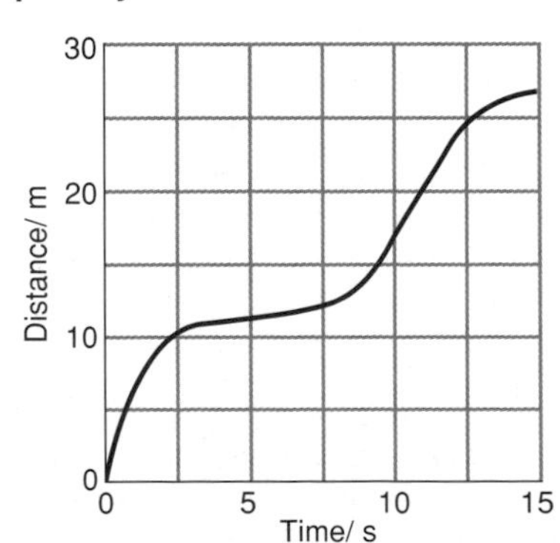

Here the unit is

$$\frac{\text{m}}{\text{s}} = \text{m s}^{-1}$$

The unit often provides a clue about whether or not the gradient has physical significance. In this case the unit shows that the gradient represents speed. Therefore the gradient of the distance–time graph is the speed.

Graphs with false origins

When a straight-line graph has a false origin the gradient can be used to determine the intercepts.

The steps are as follows:

1. Determine the coordinates (p,q) of any point on the line.
2. The y intercept is $q - \text{gradient} \times p$
3. The x intercept is 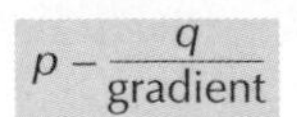 $p - \dfrac{q}{\text{gradient}}$

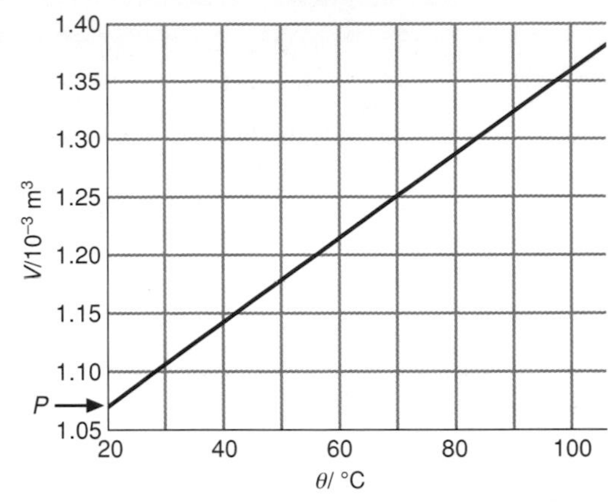

Volume–temperature graph with false origin

- Gradient $= 3.6 \times 10^{-6}\,\text{m}^3\,\text{K}^{-1}$
- Coordinates of point P $= 20\,°\text{C},\ 1.07 \times 10^{-3}\,\text{m}^3$
- Intercept on the V axis $= 1.0 \times 10^{-3}$
- Intercept on the θ axis $= -280\,°\text{C}$

Note: Temperature differences are given in K even when actual temperatures are given in °C.

Example 2: A gradient may enable the calculation of a physical constant

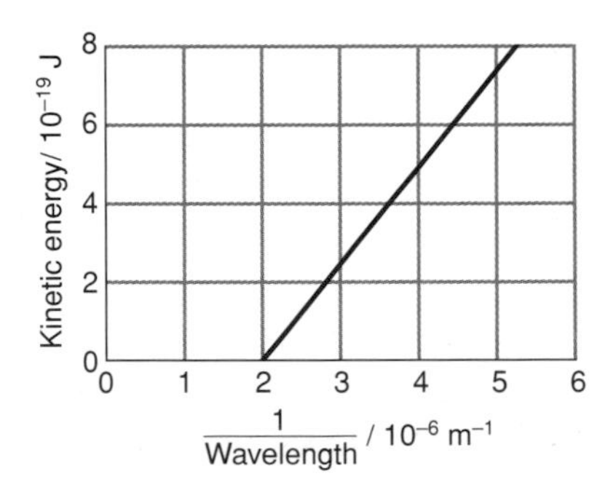

Here the gradient shows how the kinetic energy of electrons emitted from a surface in the photoelectric effect changes with 1/wavelength of the incident light. The gradient has the unit

$$\frac{\text{J}}{1/\text{m}} = \text{J m}$$

The unit does not represent any single physical quantity. In this case the gradient represents

Planck's constant × speed of light = hc

If one of these quantities is known (e.g. the speed of light) the other (Planck's constant) can be determined.

Average and instantaneous rates of change with time

When the variation of a quantity with time is not uniform, it is necessary to distinguish between the rate of change at a particular time and the average rate of change of the quantity over a period of time.

Example 1: Production of hydrogen

This graph shows how the volume of hydrogen collected varies with time during the first 300 s of an experiment.

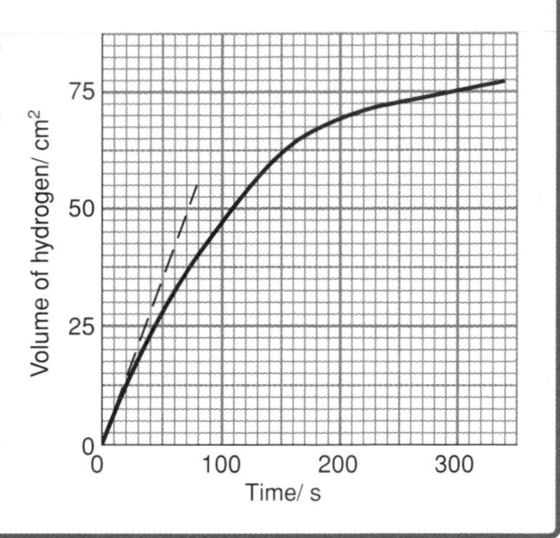

The **average** rate of production of hydrogen during the experiment is the volume produced divided by the time taken:

$$\text{Average rate} = \frac{\text{volume after 300 s}}{\text{300 s}} = \frac{75\ \text{cm}^3}{300\ \text{s}} = 0.25\ \text{cm}^3\ \text{s}^{-1}$$

The greatest **instantaneous** rate of production occurs at the start of the experiment. This is found by drawing a tangent at $t = 0$ and measuring the gradient.

$$\text{Initial production rate} = \frac{60\ \text{cm}^3}{80\ \text{s}} = 0.75\ \text{cm}^3\ \text{s}^{-1}$$

In this case the instantaneous rate of production of hydrogen decreases with time.

Example 2: A car journey

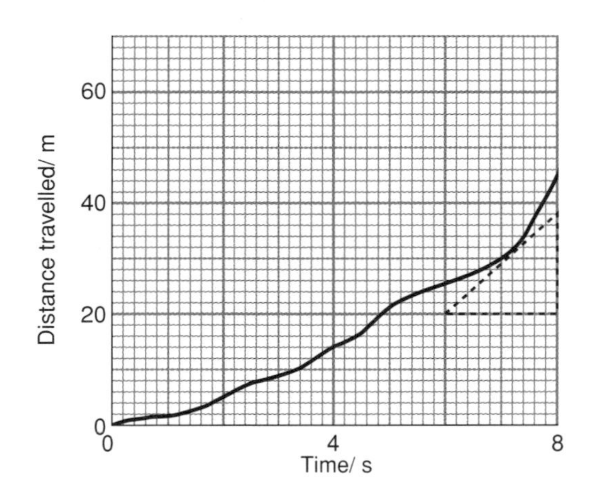

Instantaneous speed

The gradient of this graph ($\Delta s/\Delta t$) at a time T gives the speed of the car at that time. This is found by drawing a tangent to the curve at time T. The value obtained would be the reading you would see on the speedometer at that time.

A little earlier or a little later, if the driver were accelerating or braking, it would be different.

Instantaneous speed after 4.0 s = 4.1 m s^{-1}
Instantaneous speed after 8.0 s = 25 m s^{-1}

Average speed

This takes account of all the fluctuations of speed during the journey. The average speed for the part of the journey up to time T

$$= \frac{\text{total distance travelled up to time } T}{\text{total time taken } T}$$

The average speed during a given time interval is therefore found by dividing the instantaneous value of s by the time taken to travel that distance.

Average speed in the first 4.0 s = 3.5 m s^{-1}
Average speed in the first 8.0 s = 5.6 m s^{-1}

Note: When a graph is a straight line (**linear graph**) through the origin there is no difference between the instantaneous and average values. The gradient, and hence the speed, is constant at all times, so the instantaneous speed is equal to the average speed.

Note: During a journey the speed of a car varies. Sometimes it is at rest at junctions; at other times it is accelerating or travelling at constant speed. Changing gear results in changes in acceleration.

Example 2 shows a graph of distance travelled against time for a small part of the journey.

Questions to try

1. Plot a graph for a chemical reaction using the following data:

Time/10^3 s	0	1.2	2.4	4.8	7.2	9.6
Concentration /mol dm^{-3}	1.85	1.45	1.20	0.90	0.72	0.60

The instantaneous reaction rate is the gradient of the graph at any instant.

(a) Determine the initial reaction rate.
(b) Determine the reaction rate after 4.0×10^4 s.
(c) Determine the average reaction rate during the first hour.
(d) What is the unit in which the reaction rate is measured?

2. The following data relate to the mass of potassium ions taken in by a plant during a 25 minute interval.

Time/s	0	120	240	480	720	1200	1500
Mass/μg	0	20	36	49	60	73	77

(a) Use these data to plot a graph.
(b) Determine the average rate at which potassium is taken in by the plant during the first 15 minutes.
(c) Determine the initial take-up rate.
(d) Determine the take-up rate after 15 minutes.

Areas under graphs

The area between the graph line and an axis may give useful information. Sometimes the area has no meaning at all, but at other times it represents a physical quantity.

What does the area 'under a graph' mean?

This refers to the area between the graph line and the x-axis. In examples 1 and 2, this is the shaded area.

The area may or may not have a meaning. This depends on what is plotted and even then on getting the axes the right way round.

The area between the graph line and the axis represents the quantity plotted on the y-axis multiplied by the quantity on the x-axis.

You may recognize that there is a formula in which the two quantities are multiplied together to give another quantity. If so, this is the quantity that **may be** represented by the area.

Multiplying the axis units together gives the unit for the area and sometimes provides a clue to its meaning.

How is the quantity represented by the area determined?

1. *For a straight-line graph,* divide the area into triangles and rectangles or trapeziums and determine the area of each of these in cm^2.

   For a curved graph, count the number of complete 1 cm squares and estimate as accurately as you can the area represented by incomplete squares (e.g. count the number of 2 mm squares and divide by 25).
2. Calculate what 1 cm^2 of graph paper represents. Do this by multiplying the quantity represented by 1 cm on the y-axis by the quantity represented by 1 cm on the x-axis.
3. Multiply the values from steps 1 and 2 together and don't forget to give a unit.

Example 1

A graph may be plotted of *force* against *time*.

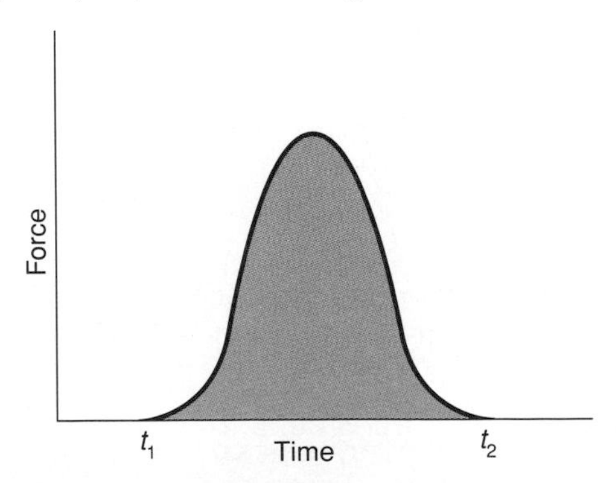

The area gives force × time, and

Impulse = force × time

so the shaded area under the graph represents the total impulse during the time from t_1 to t_2.

Example 2: A non-linear graph

Graphs of 'rate of change of a quantity with time' against 'time' are common in all the sciences. The graph shows how the rate of production of hydrogen, in $cm^3\ s^{-1}$, in a chemical reaction varies with time.

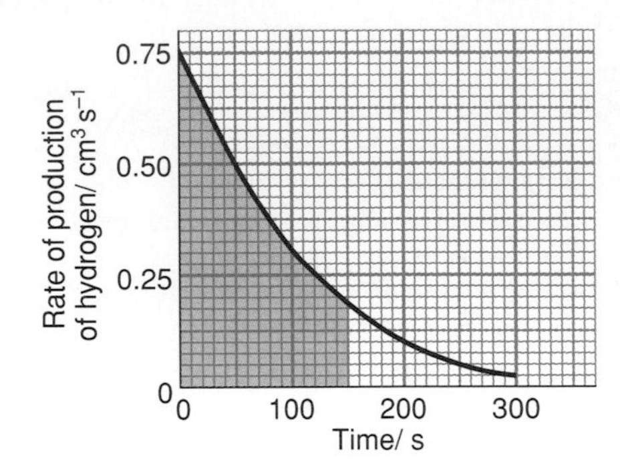

Multiplying the units on the axes together gives $(cm^3\ s^{-1}) \times s = cm^3$. The area represents volume and is the total volume of hydrogen produced in the reaction. (If the y-axis had represented rate of change of population, the area would represent the total population change in the time.)

Note that the area between the graph line and the **time** axis gives the total change. The area between the graph line and the rate of production axis also has the unit of cm^3 but this has no physical significance.

Example 3: Using a non-linear graph

Estimate the volume of hydrogen produced in the first 150 s using the graph in example 2.

Total large squares under graph from 0 to 150 s = 10 (approx.)

Each large square represents $0.125 \times 50 = 6.25\ cm^3$

Volume of gas produced = $6.25 \times 10 = 62.5\ cm^3$

This is only an approximate method, so the estimate would be best given as about 60 cm^3.

Example 4: A linear graph

The graph shows a force–extension (F–e) graph for a spring.

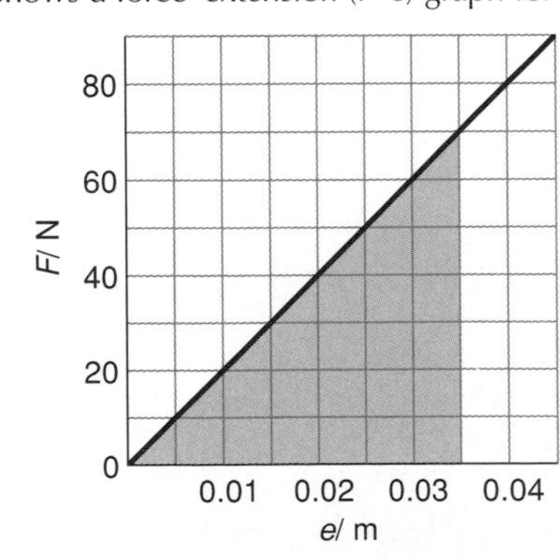

Calculate the energy stored when the extension is 0.035 m.

Force × distance = energy transformed

The area under the graph represents the stored elastic energy in the spring. The graph is linear so the area is the area of the shaded triangle:

Energy = $\frac{1}{2} \times 70 \times 0.035 = 1.23$ J

Example 5: Radioactive decay

Activity is the number of nuclei that decay per second. The area between the graph line and the time axis represents the total number of nuclei that have decayed.

Area under the graph from 0 to 2500 s ≈ 10 cm^2

One cm square represents $1 \times 10^5 \times 0.5 \times 10^3 = 5 \times 10^7$ decays

Number of nuclei decaying in 2500 s = 5.0×10^8

The average rate of decay during the 2500 s time interval is the area under the graph divided by the time. This is

$$\frac{5.0 \times 10^8}{2500} = 2.0 \times 10^5 \text{ s}^{-1}$$

Graph of activity of a radioactive sample against time

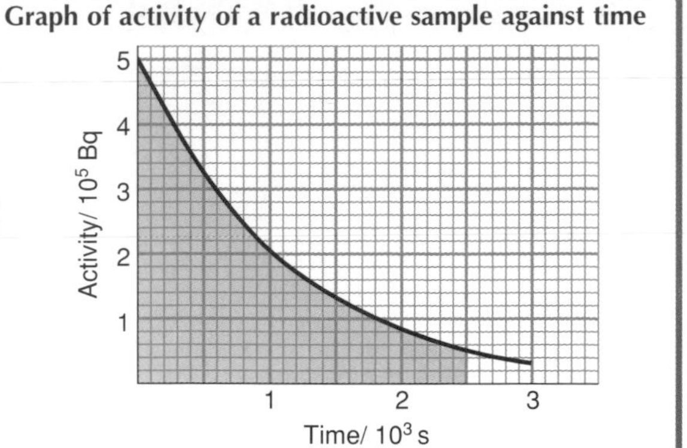

OTHER RATES OF CHANGE

Rates of change of a quantity with time are the most common, but the same principles apply to rates of change with respect to other factors.

Example 6: Graph of number of ions per mm against distance travelled by an alpha particle

The area under the graph gives the total number of ions produced by the alpha particle in the distance considered.

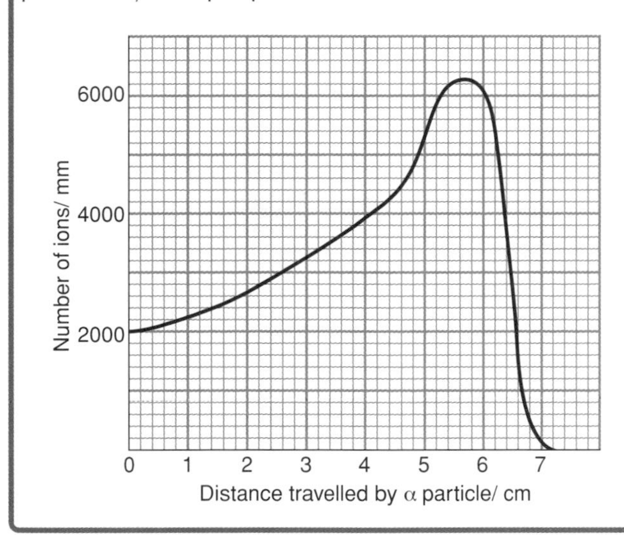

Example 7: Graph of rate of change of pressure with height

The area under the graph gives the total pressure change over the height considered.

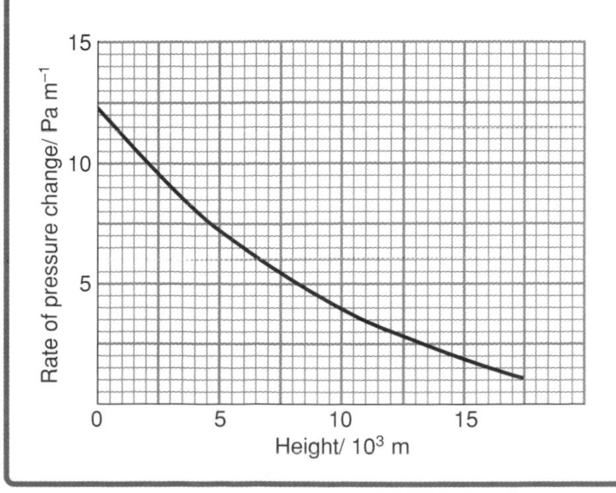

Questions to try

1. Use the graph in example 6 to determine:
   - (a) the total number of ions produced
   - (b) the percentage of the total number of ions that are produced in the last 2 cm of the range.
2. Using the graph in example 7, calculate the pressure change when rising to a height of 10 000 m.
3. What, if anything, might be represented by (a) the gradient and (b) the area under a graph of:
   - (i) force against distance travelled
   - (ii) velocity against time
   - (iii) change in population per year against the year
   - (iv) current against time.
4. The following data are for the power P developed by a sprinter against time t.

t/s	0	1	2	3	4	5
P/W	0	400	670	780	810	800

t/s	6	7	8	9	10
P/W	740	690	690	730	790

   - (a) Plot the graph.
   - (b) Determine the total energy used in the 10 s sprint.

If you want to know more about:

Triangles see page 30
Important shapes see pages 30–31

Straight-line graphs

The equation of a straight-line graph always has the form

$$y = mx + c$$

- x is the independent variable.
- y is the dependent variable.

c is the **intercept on the y-axis**. This is the value of y when $x = 0$.

Intercepts may be positive or negative.

In this case the y intercept is positive and the x intercept is negative.

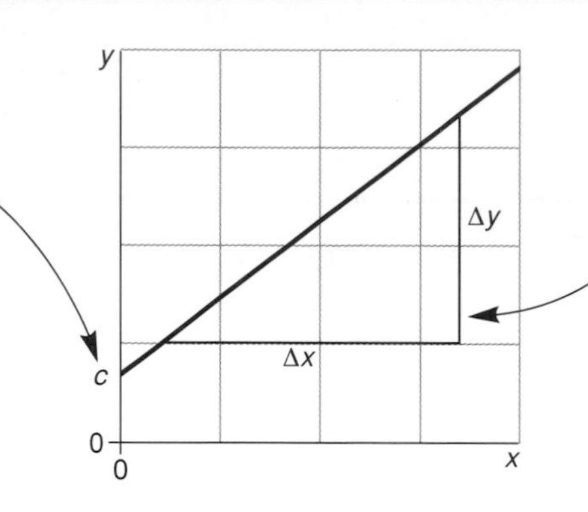

m is the gradient, $\Delta y/\Delta x$.

Gradients may be positive or negative.

In this case the gradient is positive: when x increases, y increases.

Note: When the results of an experiment produce a straight-line (linear) graph, the equation for the graph can be written down when the intercept and the gradient have been determined. The resulting equation is the **mathematical model** that relates the quantities x and y.

Much of practical physics is concerned with developing useful mathematical models. When an equation is known it is easy to predict what happens when one of the quantities is changed.

Example 1: Pressure p against temperature θ for a fixed mass of gas

From this graph we can conclude that pressure **increases linearly** with increasing temperature.

- The y intercept c is 1.2×10^5 Pa. This is the pressure of the gas at 0 °C.
- The gradient m of the graph is 4.4×10^2 Pa °C^{-1}. (Note that this would normally be written as Pa K^{-1}.)
- The equation for the line is

  $$p = 4.4 \times 10^2\theta + 1.2 \times 10^5$$

  where the unit for p is Pa (N m^{-2}) and that for θ is °C.

When $p = 0$ the equation becomes

$$0 = 4.4 \times 10^2\theta + 1.2 \times 10^5$$

θ is then $-1.2 \times 10^5/4.4 \times 10^2 \approx -273$ °C, which is an estimate of absolute zero.

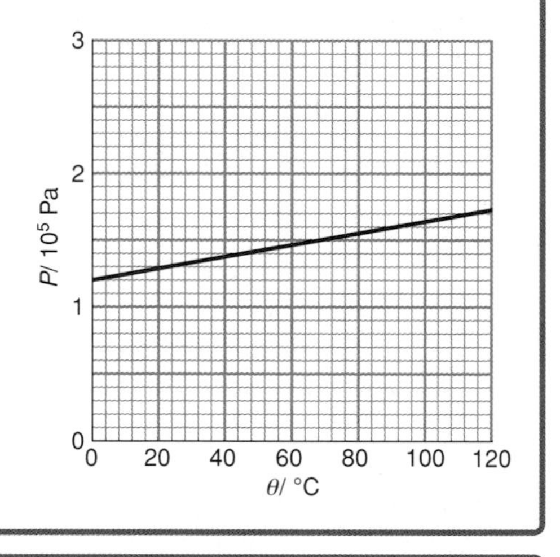

Example 2: Velocity v against time t for an object falling from rest

Since the graph is **a straight line through the origin** we can say that the velocity is **directly proportional to time**.

- The y intercept is 0. This means that when timing commenced the velocity was zero so the object was not moving.
- The gradient m of the graph is 9.8 m s^{-2}. (This is the acceleration of a freely falling object at the Earth's surface.)
- The equation for the line is

  $$v = 9.8t \quad (v = at)$$

  where the unit for v is m s^{-1} and that for t is s.

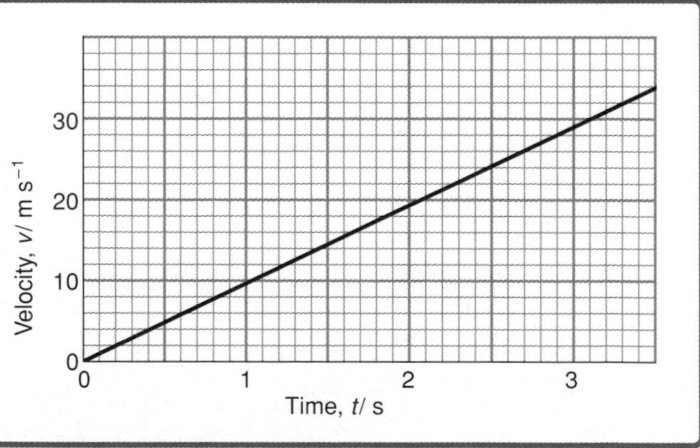

If you want to know more about:

Standard form — see pages 10–11

Units — see pages 12–13

Example 3: Terminal potential difference V against current I for a power supply

The graph has a **negative gradient**. This shows that the terminal potential difference decreases linearly with current as more current is drawn from the supply.

- The y intercept is 4.5 V. This means that when there is no current the voltage between the terminals is 4.5 V. (This is the e.m.f. of the supply.)
- The gradient m of the graph is $-3.0\ \text{V A}^{-1}$ or $-3.0\ \Omega$. (This is $-r$ where r is the internal resistance of the supply.)
- The equation for the line is

  $V = -3.0I + 4.5$ (in $y = mx + c$ form).

  This is usually quoted in the form $V = E - Ir$.

Using this equation we can predict the terminal potential difference for any current.

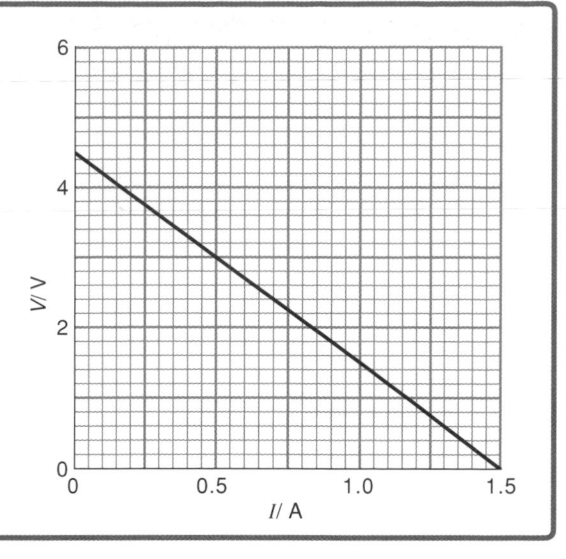

Note: The negative gradient does **not** mean that V is inversely proportional to I.

Note: In example 4, the quantity plotted on the x-axis is the reciprocal of the wavelength. This is a **derived** quantity, $1/\lambda$. The same principle is used to produce the equation relating E_K and $1/\lambda$.

Example 4: Kinetic energy E_K of emitted photoelectrons against 1/wavelength of incident radiation, $1/\lambda$

This graph shows that the **maximum kinetic energy** E_K of the electrons increases **linearly** as the frequency f of the incident radiation is increased, as frequency $\propto \frac{1}{\text{wavelength}}$.

- The intercept c on the y-axis in this case is -3.0×10^{-19} J. (This is $-\varphi$, where φ is the work function of the metal surface.)
- The gradient m is 2.0×10^{-25} J m. (This is the Planck constant multiplied by the speed of light, hc. The unit is J s × m s^{-1}.)
- The equation of the line is

  $E_K = 2.0 \times 10^{-25}\frac{1}{\lambda} - 3.0 \times 10^{-19}$.

  In symbol form the equation is

  $E_K = hc\left(\frac{1}{\lambda}\right) - \varphi$

Note that the positive x intercept means that no electrons are emitted until $\frac{1}{\lambda}$ is 1.5×10^6 because the electrons cannot have negative kinetic energy.

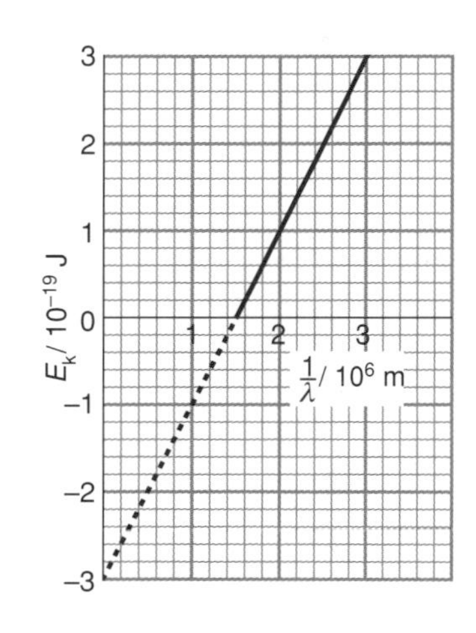

Questions to try

For each graph:

(a) Determine the y intercept and the gradient.

(b) State the physical quantity (if any) that the intercept represents.

(c) State the units and the physical quantity (if any) that the gradient represents.

(d) Write down the equation for the graph.

If you want to know more about:

Units see pages 12–13

Standard form see pages 10–11

Reciprocals see page 34

1. Distance travelled s against time t for uniform velocity

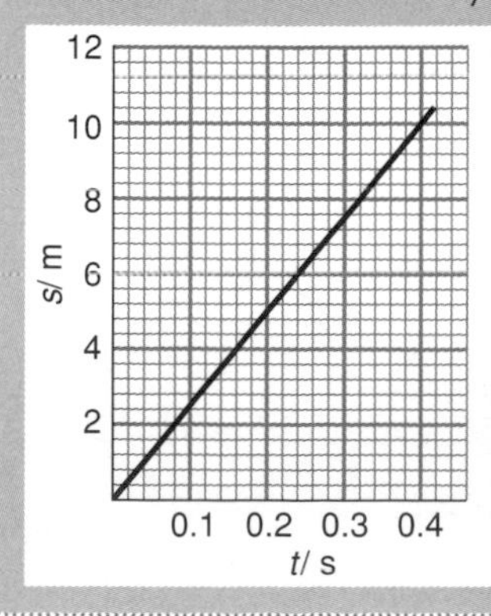

2. ΔH(vap) against boiling point θ_B

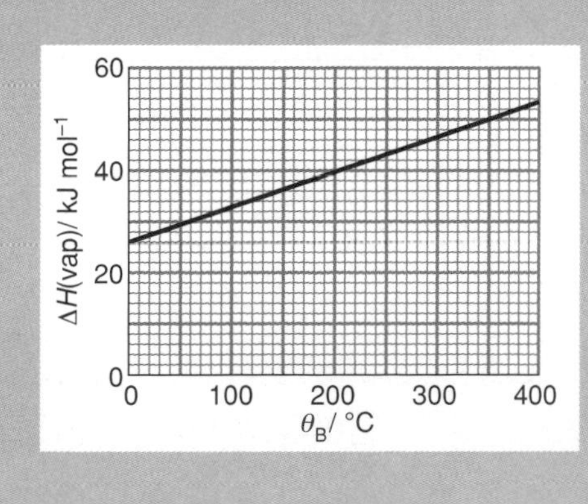

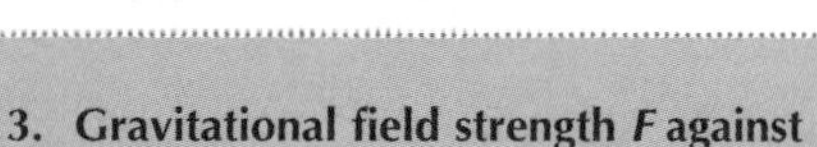

3. Gravitational field strength F against $\frac{1}{\text{distance}^2}$, $\frac{1}{r^2}$

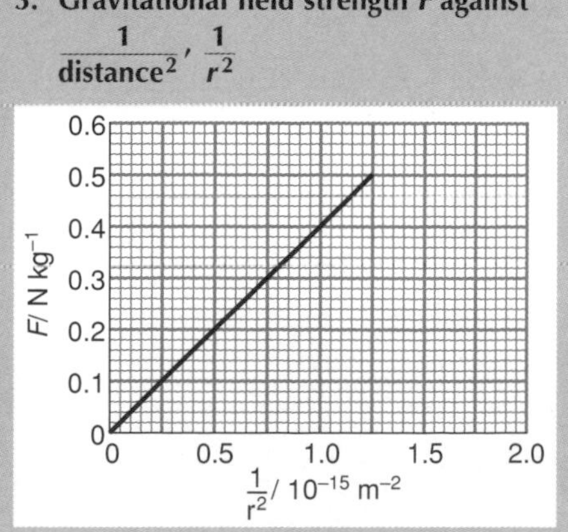

Producing straight-line graphs

To make analysis of data easier it is best to try to produce a straight-line graph whenever possible. This makes it easier to determine corresponding values of quantities between those actually measured or, by extrapolation, outside the experimental range. The constant gradient is easy to measure and often enables useful information to be found, such as physical constants. The problem is **what do you plot to get a straight line**?

General equation for a straight line

All straight-line graphs have the form

$$y = mx + c$$

The constant quantities m and c in the equation may have positive or negative values.

Graphs that go through the origin

In the special case of a graph that goes through the origin, the value of c is 0. The equation is then simply

$$y = mx$$

The majority of the equations that you will come across in your course can be converted to this form. The gradient, however, may not represent a single physical quantity but may be a compilation of a number of quantities.

Example

Stretched springs may obey Hooke's law, in which case the equation relating force F and extension e is

$$F = ke \quad (k \text{ is the spring constant})$$

Since there is no added factor there is no need to rearrange the formula.

$$\begin{array}{ccccc} F & = & k & e & + 0 \\ \Downarrow & & \Downarrow & \Downarrow & \Downarrow \\ y & = & m & x & + c \end{array}$$

The arrows indicate the corresponding quantities.

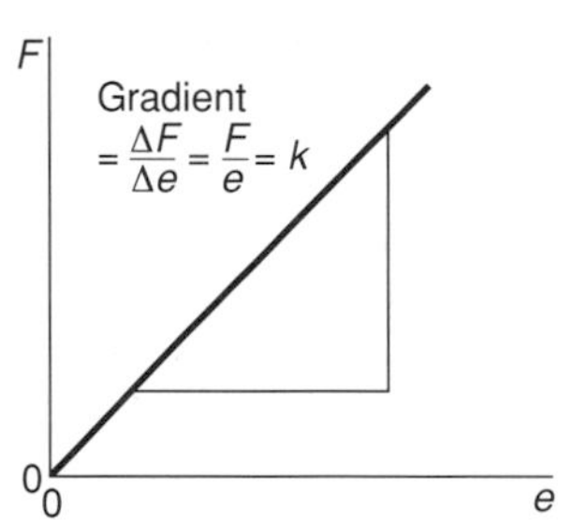

Plotting F on the y-axis and e on the x-axis gives a straight line of gradient $m = k$ that passes through the origin.

Note: The formula could have been rearranged to

$$e = \frac{F}{k}$$

or

$$\begin{array}{cccc} e & = & \frac{1}{k} & F \\ \Downarrow & & \Downarrow & \Downarrow \\ y & = & m & x \end{array}$$

Plotting e on the y-axis and F on the x-axis gives a straight line of gradient $m = 1/k$.

This shows that there is more than one way of treating the data. You just need the appropriate formula to tell you what the gradient represents.

Do you always plot measured values?

Not always. The quantities plotted on the axes may be the quantities that are measured in the experiment (see examples 1 and 2).

Often, however, a quantity **derived** from one or even both of these has to be plotted in order to obtain a linear graph. (See example 3.)

Note: Any equation with two variables and a constant term can be treated in the same way. For example:

Voltage = current × resistance

$$V = IR$$

(R is the constant m in the general formula.)

Stress = Young's modulus × strain

$$\sigma = E\varepsilon$$

In a graph of σ against ε the gradient would be E.

Questions to try

1. Measurements made are B and r. The equation relating them is
$$B = \frac{\mu NI}{r}$$
2. Measurements made are T and C. The equation relating them is
$$T = 0.69CR$$
3. Measurements made are for time taken t and measured distance d. The equation relating them is
$$v = (d + s)/t$$

In each case, state what you would plot to obtain a straight-line graph. All quantities other than those mentioned are kept constant. State what the gradient would represent and, where there is an intercept, suggest what this means.

Finding out what to plot

The task is one of rearranging the formula so that it has the form

$$y = mx + c \quad \text{or} \quad y = mx$$

It is easiest to decide what to plot on the y-axis and then change the subject of the formula to this quantity.

Example 1

The relationship between pressure, volume, temperature, and number of moles of a gas is

$$pV = nRT$$

The variation of volume with temperature of a given mass of gas was investigated at constant pressure. What should be plotted?

Rearranging gives

$$\begin{array}{ccccc} p = & \frac{nRT}{V} = & \frac{nR}{V} & T \\ \Downarrow & & \Downarrow & \Downarrow \\ y = & & m & x \end{array}$$

A graph of p against T is a straight line of gradient nR/V. If two of the quantities in the expression for the gradient are known the other can be determined.

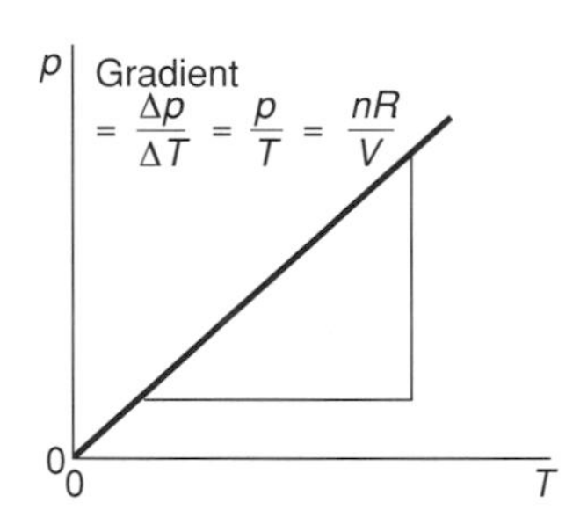

Note that T is in K in this example.

Example 2

The variation of pressure with volume was investigated for a given mass of gas at constant temperature. What should be plotted?

Rearranging the equation in example 2 gives

$$\begin{array}{ccccc} p = & \frac{nRT}{V} = & nRT & \left(\frac{1}{V}\right) \\ \Downarrow & & \Downarrow & \Downarrow \\ y & = & m & x \end{array}$$

This time the term that contains the variable V and corresponds to x is $1/V$. A graph of p against $1/V$ is a straight line of gradient nRT.

If you want to know more about:
Changing the subject of formulae see pages 24–25
Straight-line graphs see pages 42–43

Example 3: Motion of a trolley on a runway

In an experiment using light gates a cart runs down a track. The acceleration a of the cart is constant. The cart is released from the same point above the first gate so that it always has the same speed u through this gate. The distance s to the other gate is altered and the time t to reach the gate is measured. What graph could be plotted to obtain a straight line?

The equation for the motion is

$$s = ut + \frac{1}{2}at^2$$

Neither s against t nor s against t^2 will produce a straight line. Squaring or taking square roots does not help. But dividing through by t gives

$$\frac{s}{t} = u + \frac{1}{2}at$$

$$\frac{s}{t} = (\tfrac{1}{2}a)t + u$$

$$\Downarrow \quad \Downarrow \quad \Downarrow \quad \Downarrow$$

$$y = m \; x + c$$

Plotting s/t against t gives a straight line of gradient $a/2$.

The intercept on the s/t axis will be positive and represents the speed of the cart through the first gate.

Example 4: Simple pendulum

Several formulae in physics involve square roots. The simple pendulum example is used here to demonstrate how to manipulate such an equation to produce a straight-line graph and obtain useful information from it. The period of oscillation T of a simple pendulum is given by

$$T = 2\pi\sqrt{\frac{l}{g}}.$$

where l is the length of the pendulum and g is the acceleration of free fall. The data obtained in an experiment are

l/m	0.20	0.30	0.40	0.50	0.60
T/s	0.85	1.06	1.23	1.39	1.52

- What should be plotted to confirm the relationship between T and l?
- What does the gradient represent?
- The correctly drawn graph does not go through the origin as expected. What does the intercept represent?

Rearranging the formula, first rewrite the equation in the form $y = mx$:

$$T = \frac{2\pi}{\sqrt{g}}\sqrt{l}$$

A graph of T against $\sqrt{l}$ would give a straight line through the origin and would be a suitable graph to plot. However, by squaring both sides the equation becomes

$$T^2 = \frac{4\pi^2}{g}l$$

A graph of T^2 against l should be a straight line through the origin too. This is usually the preferred graph because the x-axis is the unchanged independent variable.

The gradient m is $4\pi^2/g$, so g can be found from $4\pi^2/m$.

When the experimental data given are used to plot a graph of T^2 against l the gradient is $3.9\ \mathrm{s^2\ m^{-1}}$. So

$$g = 4\pi^2/3.9 = 10\ \mathrm{m\ s^{-2}}$$

There is a small positive intercept (approximately 0.02 m) on the l axis and a small negative intercept on the T^2 axis (approximately $-0.08\ \mathrm{s^2}$). (You may wish to plot the graph accurately to verify this for yourself.) The equation for the line is therefore

$$T^2 = 3.9l - 0.08$$

When the value T is 0 there is a pendulum length. It seems that the length has to become negative for the period to become zero. The pendulum lengths have therefore been measured to be longer than they actually were. This could have been due to the use of a ruler with a large zero error or because the length was not measured to the centre of gravity of the pendulum bob.

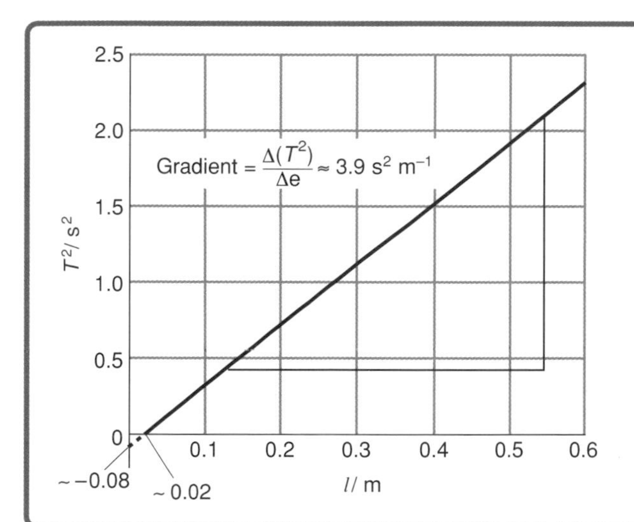

Questions to try

In each of the following investigations:

(a) what should be plotted to get a straight-line graph?

(b) what do the gradients mean?

1. An investigation of variation of the frequency f with length l of a stretched string of constant tension T and mass per unit length μ. The equation relating them is

$$f = \frac{1}{2l}\sqrt{\frac{T}{\mu}}$$

2. An investigation of how the wavelength λ associated with an electron varies with its kinetic energy E. The equation relating them is

$$\sqrt{(2mE)} = \frac{h}{\lambda}$$

where h is the Planck constant and m the mass of the electron.

Note: An intercept often appears because of a **systematic error** in the data. A systematic error is one that affects all readings equally. It may be a zero error on a meter that has not been accounted for. It may be that it is not physically possible to measure an actual distance that appears in the formula because part of the length is inaccessible. In these cases, all readings will be larger or smaller than the actual value by the same amount.

The rearranged formula enables you to interpret what the intercept means.

Error bars on graphs

No scientific measurement is absolutely accurate. Assuming that an instrument is calibrated correctly, a measurement depends on its precision, the ability of the user to read it accurately and the procedure used. Uncertainties can be minimized by taking precautions such as repeating the observation several times and avoiding introducing errors such as those due to parallax.

Assessing the uncertainty in a measurement is an important part of any experiment.

Note: Make sure that you are consistent in your data. All data should be quoted to the same number of significant figures, implying the same accuracy.

Results of an electrical experiment

Current / A ± 0.01A	Potential difference/ V ± 0.2V
0	0
0.10	1.0
0.20	2.3
0.30	3.5
0.41	4.3
0.50	5.6

Note: The uncertainty is assumed to be ± 1 in the last figure quoted unless stated otherwise.

Remember to include 0 as well as 1, 2, etc.

0.50 ± 0.01 A means that the value is between 0.49 A and 0.51 A.

2.3 ± 0.2 V means that the value is between 2.1 V and 2.5 V

Graph of potential difference across a resistor against the current through it.

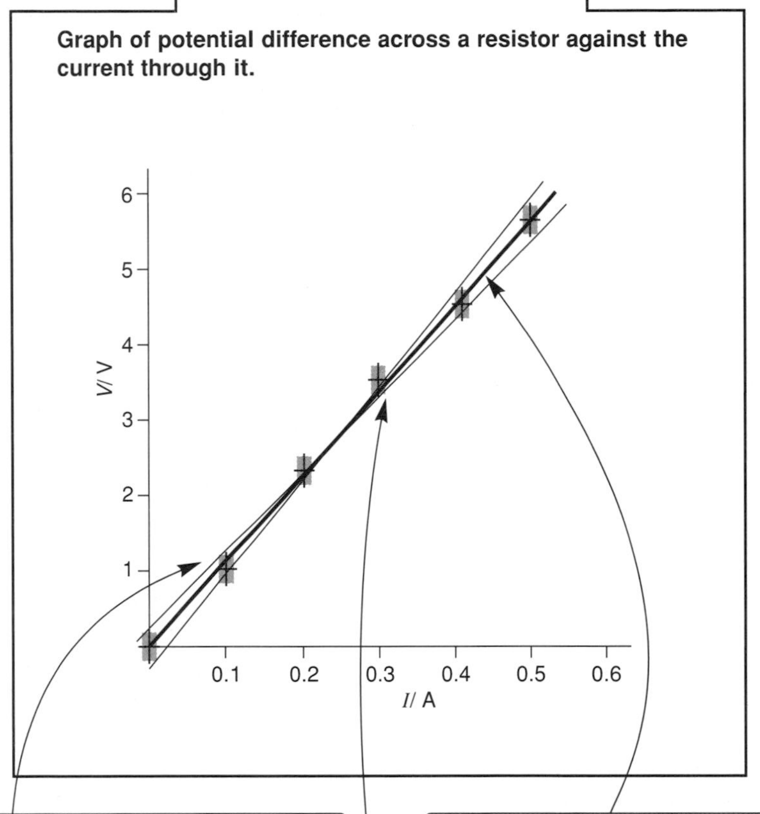

Assuming that the uncertainties have been estimated correctly the actual value will lie somewhere in this rectangle.

Only the error bars are usually included on the graph, not the complete rectangles.

The best line must pass through each of the rectangles.

If the line misses one of them by a long way it is likely that a mistake was made in the reading. If this happens do not ignore it. *Explain why* it does not fit or, best of all, *check it.*

Using the extreme lines the gradient lies between 10.2 and 12.4 V A^{-1}.

The gradient is 11.3 ± 1.1 V A^{-1}(Ω).

Note: The best line enables the most likely value for the gradient of the graph to be determined.

The gradients of the extreme lines enable the uncertainty in the gradient to be determined.

Exponential changes

There are many instances in science where one quantity varies exponentially with another. The variation may be an increase or a decrease. For example, the rate of a chemical reaction may increase exponentially with temperature, and the intensity of gamma radiation may decrease exponentially with the thickness of absorbing material between the source and a detector.

Quantities that vary exponentially with time are common in advanced-level work. For example, the rate of growth of bacteria can increase exponentially with time, and the rate at which a capacitor loses its charge decreases exponentially.

An exponential change occurs when the rate of change of something at any time is proportional to how much of it there is.

The change may produce a rise or fall in the size of the quantity.

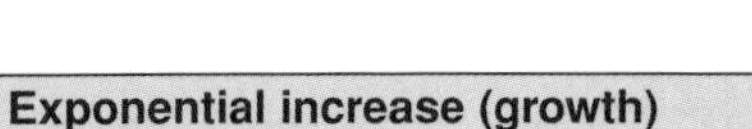

Exponential increase (growth)

When an infectious disease goes unchecked, as more people have the disease, more will catch it. The rate at which the disease spreads increases. The number of people with the disease would increase exponentially.

If 10 people have the disease and every person with the disease passes it on to three others each week, the number of people with the disease each week, and the number catching it, are:

Week	1	2	3	4	5
Number with disease	10	40	160	640	2560
Number catching disease	30	120	480	1920	7680

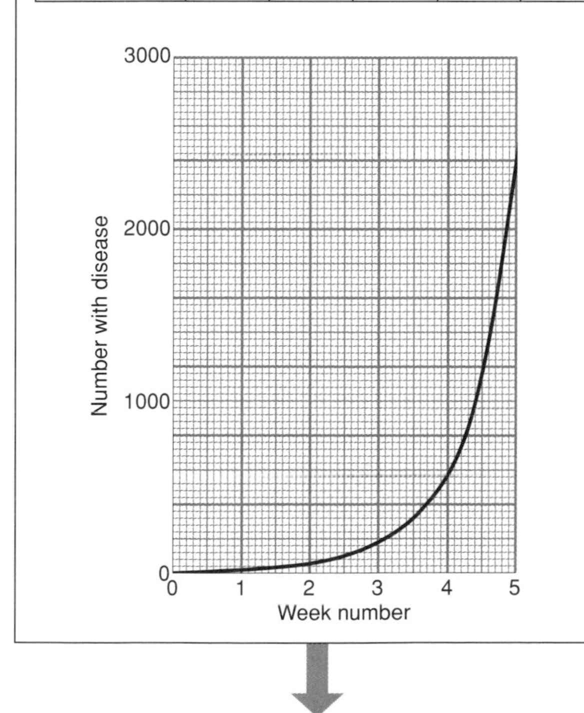

Question to try

Plot a graph to show that the rate at which people catch the disease with time is similar to the number with the disease against time.

Exponential decrease (decay)

When an infectious disease is treated, fewer people carry the disease. The rate at which the disease spreads to those not infected therefore slows down and the number with the disease gradually decreases.

Suppose a treatment for a disease has an approximately 30% success rate each week. This means that about three of every 10 people with the disease are cured each week. The number of people with the disease and the number cured varies week by week:

Week	1	2	3	4	5
Number with disease	2560	1792	1254	878	615
Number cured	768	538	376	263	185

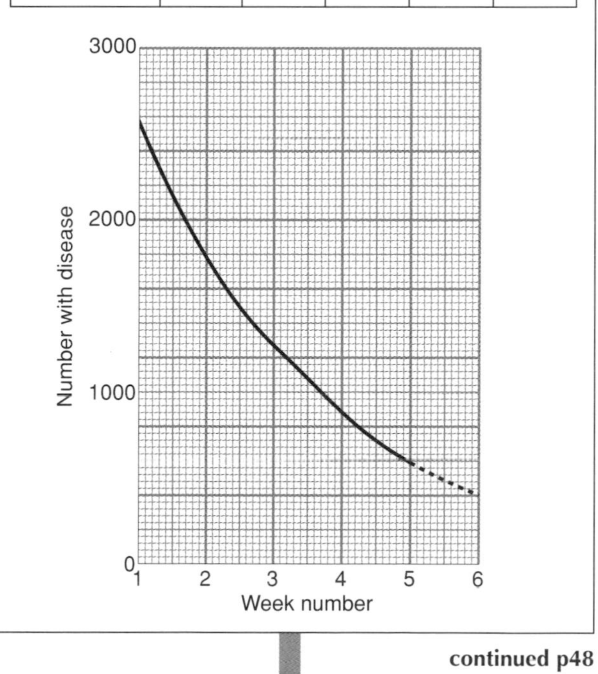

continued p48

Question to try

Plot a graph to show that the rate at which people are cured against time is similar to the number with the disease against time.

Time constant

The product RC is the time for the charge to fall to 1/e of the initial value and is called the time constant τ of the circuit.

$1/e = 1/2.72 = 0.368$

In RC seconds the charge falls to 37% of the initial value. This is shown below.

When $t = RC$,

$$Q = Q_0 e^{-RC/RC} = Q_0 e^{-1} = \frac{Q_0}{e}$$

Example: Capacitor discharge

$$\frac{\Delta Q}{\Delta t} = -\text{ constant} \times Q$$

Q is the charge on the capacitor. The variation of Q with time is

$$Q = Q_0 e^{-kt}$$

Q_0 is the charge when timing commences, i.e. $t = 0$.

The rate at which a capacitor loses charge depends on how much charge there is on it.

- In capacitor discharge the constant k is

$$\frac{1}{\text{resistance} \times \text{capacitance}} \quad \left(\frac{1}{RC}\right)$$

RC is referred to as the **time constant**, τ.

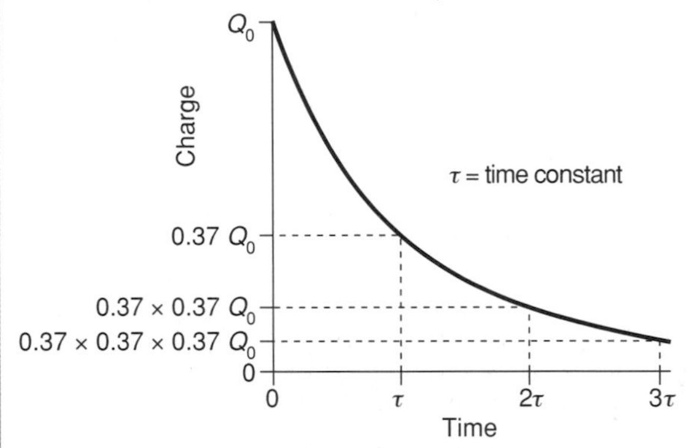

Note: In the equations e is the exponential constant. This has a value of 2.718 to 4 significant figures.

Relationship between time to halve and the constant *k*

$$\frac{N}{2} = Ne^{-kt_{1/2}}$$

Cancelling N on each side and taking the ln of each side gives

$$\ln(\tfrac{1}{2}) = \ln(e^{-kt_{1/2}}) = -kt_{1/2}$$

$\ln(\frac{1}{2}) = \ln 0.5 = -0.69$, so

$$-0.69 = -kt_{1/2}$$

Rearranging the equation gives

$$t_{1/2} = \frac{0.69}{k}$$

For capacitor discharge,

$$\text{Time to halve} = 0.69RC$$

For radioactive decay,

$$\text{half-life} = \frac{0.69}{\lambda}$$

Half-life

The time for the number of radioactive nuclei of the isotope, or for its rate of decay (its activity), to halve is referred to as the **half-life**.

Example: Radioactive decay

$$\frac{\Delta N}{\Delta t} = -\text{ constant} \times N$$

The variation of N with t is $N = N_0 e^{-kt}$

N_0 is the number of radioactive atoms when timing commences, i.e. $t = 0$.

- The rate at which atoms of a radioactive isotope decay depends on how many radioactive atoms of the isotope there are.
- In radioactive decay the constant k is called the decay constant. It is given the symbol λ.

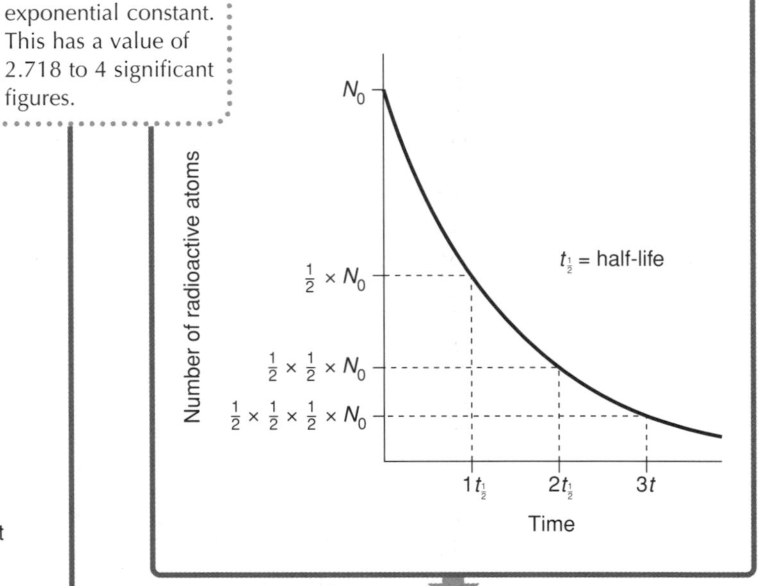

Change of activity in radioactive decay

- After one half-life the activity is $\frac{1}{2}$ that of the original.
- After 2 half-lives it is $\frac{1}{2} \times \frac{1}{2} = \frac{1}{4}$.
- After three half-lives it is $\frac{1}{2} \times \frac{1}{2} \times \frac{1}{2} = \frac{1}{8}$.
- After n half-lives the activity is $(\frac{1}{2})^n$.

Note that when the decay goes on for a long time the value will approach, but never reach, zero.

The same happens for a given capacitor discharging through a given resistor. The capacitor always takes the same time to lose half its charge.

Note: Test for exponential decay

The time for the quantity to halve is a constant.

- Read the value at any time.
- Calculate half that value.
- Determine from the graph the time taken to halve.
- Repeat for other starting values.
- If the graph is exponential the time should always be the same.

Note that the time taken to fall by or to any given fraction ($\frac{1}{2}, \frac{1}{3}, \frac{1}{4}$, etc.) is always the same.

Using the decay equations

Example 1: Calculating the value after a given time
The number of radioactive atoms in a sample of strontium-90 is 1.60×10^{20} at time $t = 0$. The half-life is 28 years. How many atoms are left after (a) 84 years (b) 300 years?

First look at whether the time is a whole number of half-lives.

(a) In this case 84 years = 3 half-lives.
The number of radioactive atoms left is $\left(\frac{1}{2}\right)^3$ of the original number:

$$= \frac{1}{8} \times 1.6 \times 10^{20} = 0.20 \times 10^{20} = 2.0 \times 10^{19}$$

(b) In this case 300 years = 10.7 half-lives. There are two possible methods.

Method 1

Number of radioactive atoms left

$$= \left(\tfrac{1}{2}\right)^{10.7} \text{ of the original number}$$
$$= \left(\tfrac{1}{2}\right)^{10.7} \times (1.6 \times 10^{20})$$

Method 2

Determine the decay constant using $\frac{0.69}{\text{half-life}}$

Decay constant $\lambda = 0.0246$ years^{-1}

(Note that as long as all times are in years there is no need to change to seconds.)

$$\text{Number left, } N = N_0 \, e^{-\lambda t} = 1.6 \times 10^{20} e^{-0.0246 \times 300}$$
$$= 1.6 \times 10^{20} \, e^{-7.38}$$
$$= 1.6 \times 10^{20} \times 6.24 \times 10^{-4}$$
$$= 1.0 \times 10^{17}$$

To calculate $e^{-7.38}$:

- Enter –7.38 into your calculator using the special +/– key to make the number negative.
- Press the e^x key (usually INV followed by ln).
- Press =.
- To complete the sum multiply the answer by 1.6×10^{20}.

Example 2: Calculating the rate of change for a given value of the quantity
The half-life of potassium-42 is 4.5×10^4 s. What will be the rate at which atoms disintegrate when a sample contains 5.0×10^{10} atoms of potassium-42?

$$\frac{\Delta N}{\Delta t} = -\lambda N$$

$$\lambda = \frac{0.69}{\text{half-life}} = \frac{0.69}{4.5 \times 10^4 \text{ s}} = 1.53 \times 10^{-5} \text{ s}^{-1}$$

$$\frac{\Delta N}{\Delta t} = -(1.53 \times 10^{-5}) \times (5.0 \times 10^{10}) = -7.65 \times 10^5 \text{ s}^{-1}$$

The negative sign means that the number of radioactive potassium atoms is decreasing.

Example 3: Calculating the time to reach a given value
A capacitor has a charge of 2000 μC. It discharges through a resistor. The time constant, RC, of the discharge is 30 s. After what time will the charge be (a) 500 μC (b) 1400 μC?

First check to see whether the charge can be achieved exactly by halving then halving again and so on. The time is then equal to the number of times halving has occurred multiplied by $0.69RC$.

(a) In this case half of 2000 μC = 1000 μC
half of 1000 μC = 500 μC

The charge has to be halved twice so the time required is

$$2 \times 0.69 \times 30 = 41 \text{ s}$$

(b) In this case the charge does not halve so the time is less than $0.69RC$.

$$Q = Q_0 e^{-t/RC}$$
$$1400 \ \mu\text{C} = (2000 \ \mu\text{C}) e^{-t/30}$$
$$\frac{1400 \ \mu\text{C}}{2000 \ \mu\text{C}} = 0.70 = e^{-t/30}$$

Take ln (log to base e) on both sides.

Note that $\ln(e^{-t/30}) = -\frac{t}{30}$.

$$-0.357 = -\frac{t}{30}$$

$$t = 30 \times 0.357 = 10.75$$

Questions to try

1. The half-life of free neutrons is 10.8 min. Calculate the decay constant of neutrons in s^{-1}.
2. How long will it take the activity of a sample of iodine-135 (half-life 6.7 h) to fall to:
   (a) $\frac{1}{16}$ of its initial value?
   (b) $\frac{1}{20}$ of its initial value?
3. A sample of radium of half-life 1600 years contains 5.5×10^{20} radioactive atoms.
   (a) How long will it take for the number of radioactive atoms to become 3.2×10^{18}?
   (b) How many radioactive atoms will there be after 500 years?
   (c) Calculate the initial activity of the sample.
4. The time constant for the discharge of a capacitor through a resistor is 25 s.
   (a) Calculate the current in the circuit when the capacitor has a charge of 1500 μC.
   (b) Determine the charge on the capacitor when the current is 15 μA.

If you want to know more about:
Standard form see pages 10–11
Using your calculator see pages 26–27
Exponential changes see pages 47–48

Charging a capacitor / Exponential growth

This is more difficult to understand than the discharge graph (page 48) and it is advisable to become familiar with the discharge graph first. The charging process is particularly important to students following electronics courses because both capacitor charging and discharging are used in timing circuits.

The charging graph

The graph shows how the charge Q on a capacitor varies with time t as it is charged through a resistor. The graph of voltage against time has the same shape ($Q \propto t$).

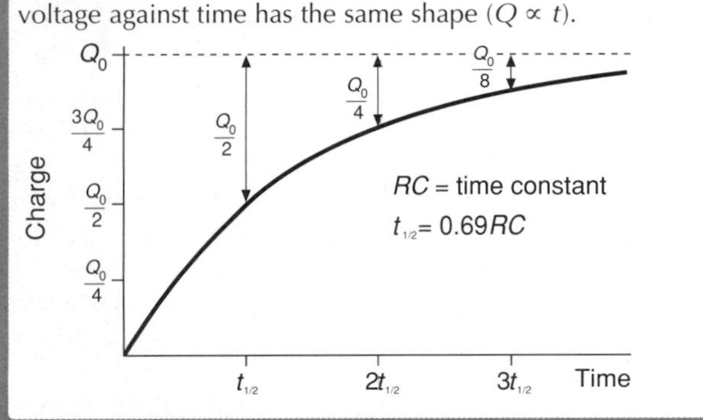

Features of the charging graph

- The rate of charging decreases as the capacitor charges; this is shown by the decreasing gradient.
- If the maximum charge is Q_0 the charge reaches $0.63Q_0$ in a time equal to the time constant (RC). Note that $0.63 = (1 - 1/e)$.
- The charge is

  $\frac{Q_0}{2}$ below the maximum after time $t_{1/2} = 0.69RC$

  $\frac{Q_0}{4}$ below the maximum after time $2t_{1/2}$

  $\frac{Q_0}{8}$ below the maximum after time $3t_{1/2}$

  and so on.
- The charge approaches the maximum Q_0 but never reaches it.

For practical purposes the capacitor is assumed to be fully charged after time 5*RC*.

The charging equation

The charging curve is an upside-down version of the discharging curve.

The equation for a charging curve is

$$Q = Q_0 - Q_0 e^{-t/RC} = Q_0(1 - e^{-t/RC})$$

In terms of the voltage across the capacitor, which is usually more useful in practical work, the equation is

$$V = V_0 - V_0 e^{-t/RC} = V_0(1 - e^{-t/RC})$$

Example: Using the equation

The time constant of a circuit is 2.5 ms. Calculate the time for a capacitor to reach 75% of its maximum voltage.

The fact that the voltage V reaches 75% of its maximum means that $V/V_0 = 75/100 = 0.75$.

Although it is not essential, it is safer to work with t in seconds: 2.5 ms = 0.0025 s.

From the equation,

$$V/V_0 = (1 - e^{-t/RC})$$
$$0.75 = 1 - e^{-t/0.0025}$$
$$0.75 - 1 = -e^{-t/0.0025}$$
$$0.25 = e^{-t/0.0025}$$

Taking the ln of both sides:

$$\ln 0.25 = -t/0.0025$$
$$-1.39 = -t/0.0025$$
$$t = 1.39 \times 0.0025 = 3.5 \times 10^{-3}\ \text{s}$$

If you want to know more about:

Percentages see page 18

Logarithms see pages 65–66

Equation for exponential growth

$$\frac{\Delta N}{\Delta t} = \text{constant} \times N \quad \text{and} \quad N = N_0 e^{kt}$$

- N_0 is the initial value at time $t = 0$.
- N is the value after time t.

Example: Growth of a yeast cell population

Number of cells: 0, 10, 20, 30, 40, 50, 60
Time/ h: 0, 10, 20, 30
2 cells present at $t = 0$

The graph assumes that each yeast cell divides into two cells once in a 4-hour period. In practice the growth depends on factors that change with time, so that real curves will not be true exponential curves.

Features of the growth graph

- As time increases the number of cells present increases, so the rate at which they increase becomes greater. That is, the number increases at a faster rate.
- It always takes the same time for the number to double.

Both the quantity concerned and the rate of increase vary exponentially.

Test for exponential growth

This is similar to that for exponential decay:

- Read the value at any time.
- Calculate twice that value.
- Determine from the graph the time taken to double.
- Repeat for other starting values.
- If the graph is exponential the time should always be the same.

Inverse and inverse square laws

Inverse relationships are those in which a quantity is proportional to $1/r^n$, where n is a positive number. The number can take any value but the important inverse laws that you will meet are the simple inverse relationship where a quantity is proportional to the reciprocal of another ($n = 1$) and the inverse square law in which $n = 2$.

What is an inverse square law?

When light energy radiates uniformly in all directions from a **point source** the intensity, that is the energy falling on a square metre of a surface, decreases with distance.

This decrease obeys an inverse square law: the intensity is inversely proportional to the square of the distance.

$$\text{Intensity} \propto \frac{1}{r^2}$$

The graph shows how the photon intensity (number per m^2 per s) will vary with distance when the number of photons emitted by a point source each second is 8.0×10^{20}.

Photon intensity/ 10^{20} m^{-2} s^{-1}

Distance from source/ m

Why do inverse square laws occur?

Inverse square laws occur because something spreads out from a point so that at a given distance from the source it is distributed over the surface of a sphere. Energy is a common example and will be considered here.

r

N photons

Photons spread out over area $4\pi r^2$

In the case of light the energy is carried by photons. The energy emitted per second is proportional to the number of photons emitted per second.

Suppose there were N photons emitted by the source each second. The area through which they pass at a distance r from the source is $4\pi r^2$. The number passing through a square metre is

$$\frac{N}{4\pi r^2}$$

Since N, 4 and π are all constant, the number of photons per m^2 and hence the energy intensity is proportional to $1/r^2$.

Consequence of an inverse square law

The intensity is decreased by a quarter when the distance from the source doubles. When the distance is trebled the intensity falls to 1/9 of the original intensity, and so on.

What quantities are related by inverse square laws?

The inverse square law is obeyed by all parts of the electromagnetic spectrum provided there is no absorption by the medium in which it travels.

The inverse square law also appears in work on gravitational and electric fields. In this case the variation of field strength with distance obeys an inverse square law:

$$\text{force per kg} = GMm/r^2$$

where M and m are masses and G is the universal gravitational constant.

Note: The field of a **gravitational/electric field** can be envisaged as field lines radiating outwards from a point mass/charge. The number of lines passing through a m^2 represents the strength of the field and this leads to the inverse square law.

As with energy, the further you are from the source the greater the area over which the field lines are distributed.

Inverse *r* laws

When one quantity V is proportional to the reciprocal of another quantity r, the quantities are said to obey an inverse relationship. In this case

$$V = \frac{\text{constant}}{r}$$

In this case when r is doubled V is halved, when r is trebled V is reduced to 1/3 of the initial value, and so on. Notice that the relationship works both ways.

Important phenomena that obey an inverse r law are the potential energy due to point masses and point charges. For example:

$$\text{Electrical potential energy} = kQq/r$$

where Q and q are charges and k is a constant.

Sketching graphs for inverse *r* and inverse square laws

The general curvature of the graphs for $1/r$, $1/r^2$ and an exponential is similar but the detail is quite different. For the same change in distance an exponential graph would change by the same fraction each time.

When asked to sketch graphs make sure that the sketch clearly shows the correct variation.

Quantity expressed as a percentage

$\frac{1}{r}$

$\frac{1}{r^2}$

Distance r/ m

Question to try

The force on an electron due to a proton obeys an inverse square law for force. The force is 2.3×10^{-8} N when the separation is 1.0×10^{-10} m.

(a) What is the force when the separation is 3.0×10^{-10} m?

(b) At what separation will the force be 1.0×10^{-9} N?

If you want to know more about:

Standard form see pages 10–11
Exponential changes see pages 47–48

Non-graphical testing of data

You do not have to draw a graph to check whether data might agree with a particular law. You can check whether data obey a given law or, in an investigation, perform an initial test on data by means of simple calculations. To draw a reasonable conclusion you need at least three sets of data.

General principle of the test

If a quantity y is directly proportional to x^n, a calculation of y/x^n is always the same. Note that when using practical data the value will be only approximately constant because the data will contain experimental errors.

The laws to be tested fall into one of three categories: **direct proportionality**, **inverse proportionality** and **exponential relationships**.

Testing for direct proportionality

When the y quantity seems to be increasing at a greater rate than the x quantity, the relationship could be a direct proportionality or an exponential growth.

When the rate of growth is decreasing, the y quantity could be directly proportional to the square root or cube root of the x quantity.

Examples of direct proportionality are:

- $y \propto x$, in which case y/x is constant
- $y \propto x^2$, in which case y/x^2 is constant
- $y \propto x^{1/2}$, in which case $y/x^{1/2}$ is constant.

Example: Chemical reaction

The data show how the volume of carbon dioxide produced in a reaction between calcium carbonate and hydrochloric acid varied with time. Do the data suggest that the volume V produced is proportional to the time t for which the reaction is taking place?

Time / s	0	10	20	30	40
Volume / cm³	0	23	45	70	85

If $V \propto t$ it follows that $V = \text{constant} \times t$. Therefore

$\frac{V}{t}$ should be constant

Dividing each volume by each time (ignoring $t = 0$) gives

2.30   2.25   2.33   2.12

Allowing for the uncertainties in the experimental data, the first three values are similar so there is good evidence that the volume is proportional to the time during the first 30 s. The value after 40 s is much lower than the others and suggests that the volume is then no longer proportional to time.

Testing for inverse proportionality

When the y quantity decreases as the x quantity increases, the relation could be an inverse power relationship or an exponential decay.

Examples of inverse proportionality are:

- $y \propto 1/x$, in which case xy is constant
- $y \propto 1/x^2$, in which case yx^2 is constant
- $y \propto 1/x^{1/2}$, in which case $yx^{1/2}$ is constant.

Example: Gravitational field

The following data were obtained for the gravitational field strength g at different distances from the centre of the Earth r using a space probe. Do these data confirm that the gravitational field strength follows an inverse square law?

r/ 10^7 m	5.2	4.2	3.3
g/ N kg⁻¹	0.15	0.23	0.37

An inverse square law means that

$$g \propto \frac{1}{r^2} \quad \text{so that} \quad g = \frac{\text{constant}}{r^2} \quad \text{and} \quad gr^2 = \text{constant}$$

The constant for each set of data is

4.1×10^{14} 4.1×10^{14} 4.0×10^{14}

Allowing for the uncertainties in each set of data, gr^2 leads to the same number, so the data are consistent with an inverse square law.

Testing for exponential change

When a change is **exponential** for the same change on the x-axis (e.g. the same time interval) the ratio

$$\frac{\text{new value of y}}{\text{old value of y}}$$

will be constant.

Example: Plant growth

The following data were obtained during an investigation of the effect of temperature on the growth rate of excised roots of a sweet pea plant. Do the data suggest that the effect of temperature on the growth rate is exponential?

Temperature/ °C	9	12	15	18
Growth rate/ mm per day	10	13	15	19

For the relationship to be exponential

$$\frac{\text{growth rate after a 3 °C rise}}{\text{original growth rate}} = \text{constant}$$

From 9 °C to 12 °C the ratio is 13/10 = 1.3. The other ratios are 15/13 = 1.2 and 19/15 = 1.3. Within the limits of uncertainty in the data, this suggests that there is an approximately exponential effect over this temperature range that would be worth exploring further.

Question to try

The following data were obtained for the value of capacitance C in the circuit and the resonant frequency f of a tuned circuit.

C/μF	1.0	10	100
f/kHz	500	160	50

Determine which of the following relationships is correct for this data: $f \propto C^{-1}$ $f \propto C^{-2}$ $f \propto C^{-0.5}$

Properties of triangles

General properties of any triangle

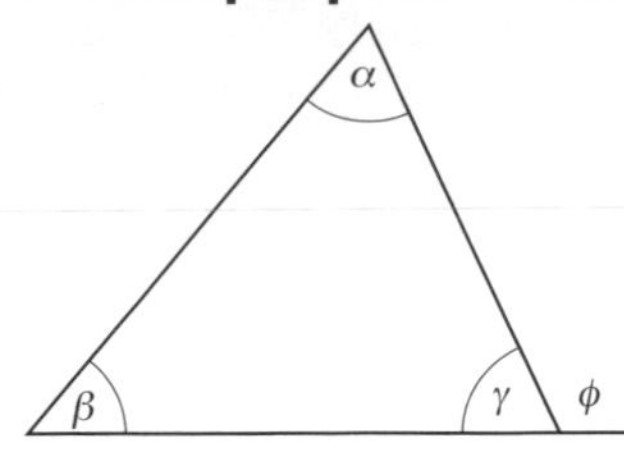

- The sum of internal angles is 180°:

  $\alpha + \beta + \gamma = 180°$

- The exterior angle equals the sum of the two internal opposite angles:

  $\phi = \alpha + \beta$

- Area $= \frac{1}{2}$ base × height $= \frac{1}{2}\,bh$

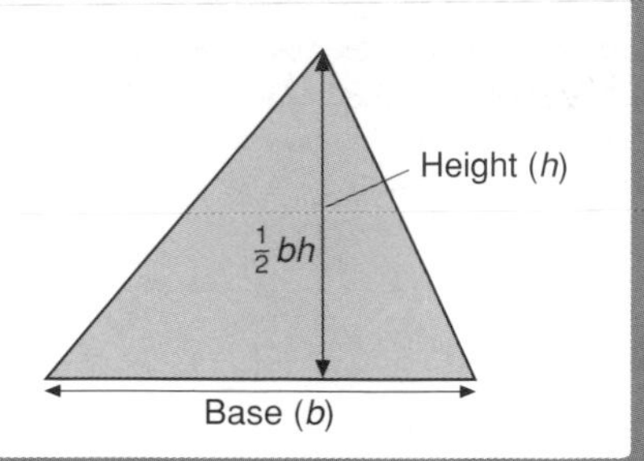

Special triangles

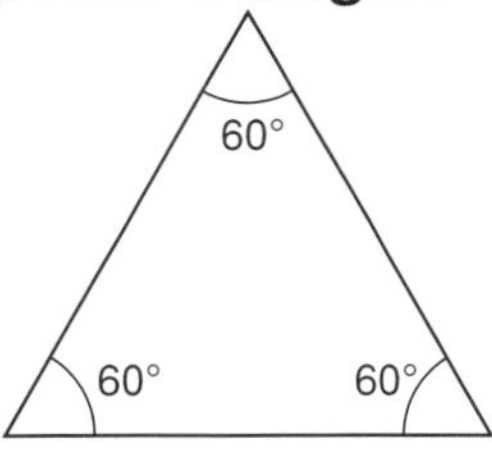

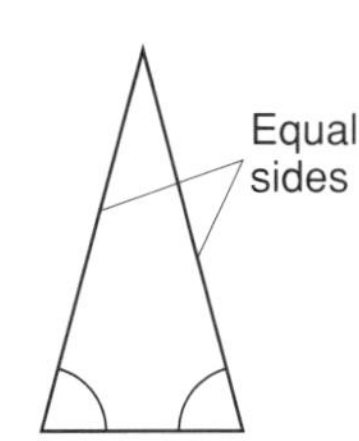

An equilateral triangle
- All angles are equal (60°).
- All sides are equal.

An isosceles triangle
- Two sides are equal.
- Two angles are equal.

Similar triangles
- They both have the same angles.
- The ratios of the sides are the same.

Right-angled triangles

The properties of the right-angled triangle form the basis of trigonometry. An understanding of the sines, cosines and tangents is necessary to do calculations with vector quantities. Pythagoras's theorem (page 54) also applies to a right-angled triangle.

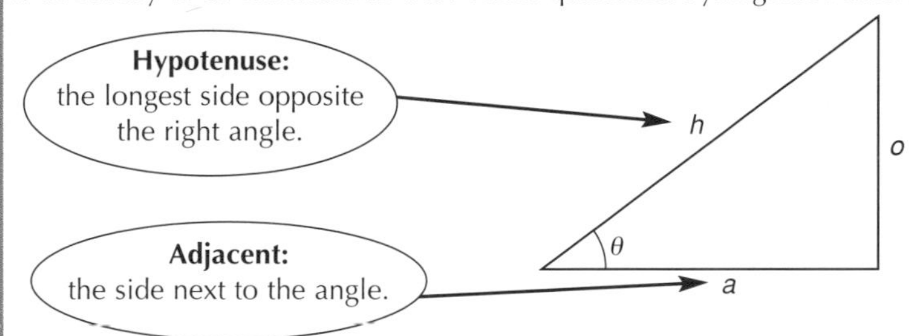

Opposite: the side opposite the angle.

Note: Be prepared to see right-angled triangles in different orientations on diagrams. Turn the page if it makes it easier for you to relate the sides to the sine, cosine and tangent.

Sines, cosines, and tangents

$$\sin\theta = \frac{\text{opposite}}{\text{hypotenuse}} = \frac{o}{h}$$

$$\cos\theta = \frac{\text{adjacent}}{\text{hypotenuse}} = \frac{a}{h}$$

$$\tan\theta = \frac{\text{opposite}}{\text{adjacent}} = \frac{o}{a}$$

These equations can be used to find:
- an angle when two sides are known
- a side of the triangle if one angle and one side are known.

Example

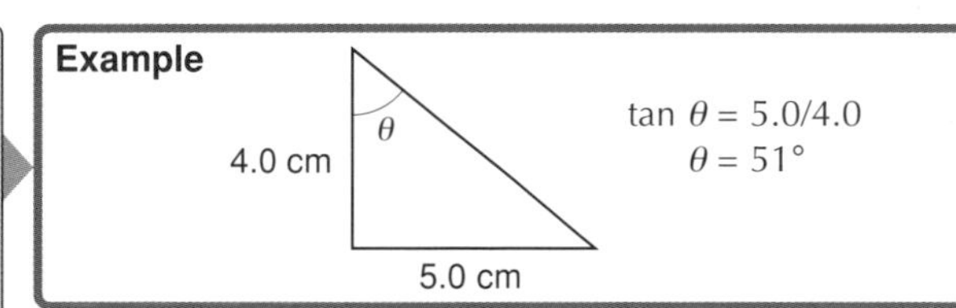

$\tan\theta = 5.0/4.0$

$\theta = 51°$

Example: Sides represent forces

6.0 (N)/F = sin 35

$$F = \frac{6.0\text{ N}}{\sin 35} = \frac{6.0}{0.57}$$

$= 10.5$ N

The sides of the triangle may represent distances (displacements). They may also represent the magnitude (size) and direction of vector quantities, e.g. force, velocity, momentum and electric field strength.

Questions to try

1. Determine the sine, cosine and tangent of the angle marked in each of the following triangles:

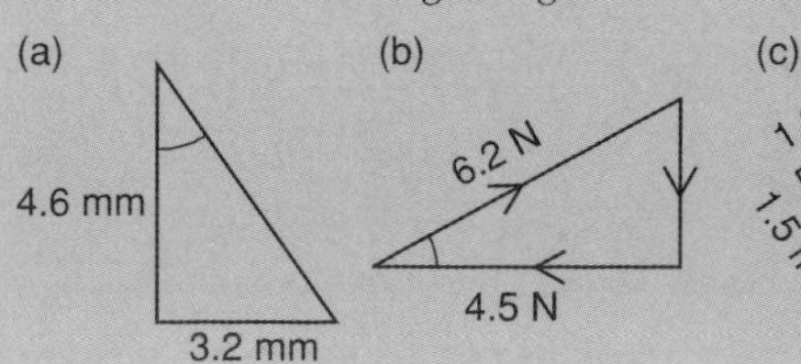

2. Calculate the unknown angles and values corresponding to the sides of the triangles below:

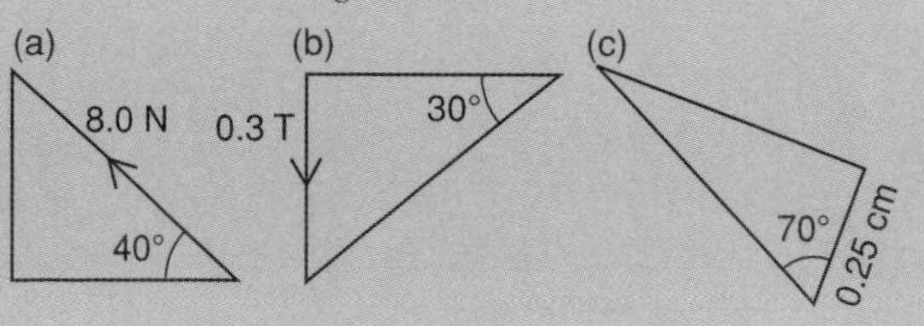

Pythagoras' theorem

This famous theorem applies to any right-angled triangle.

Pythagoras' theorem

The hypotenuse squared is the sum of the squares of the other two sides.

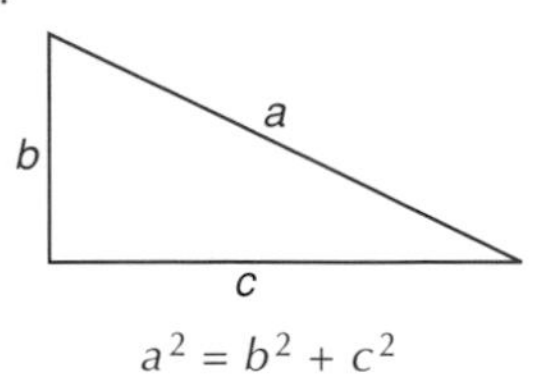

$$a^2 = b^2 + c^2$$

If two sides are known the other can be calculated.

Note: Any triangle with sides in the ratio 3 : 4 : 5 is a right-angled triangle, e.g.

$5^2 = 4^2 + 3^2$

$0.5^2 = 0.4^2 + 0.3^2$ etc.

Example 1

An environmental scientist investigating whether strong magnetic fields affect plant growth wants to know how far a ground plant is from a high-voltage cable. The cable is known to be 30 m above the ground. The horizontal distance from the plant to the cable is 25 m. How far is the plant from the cable?

$$\begin{aligned} d^2 &= (30^2 + 25^2)\ \text{m}^2 \\ &= (900 + 625)\ \text{m}^2 \\ &= 1525\ \text{m}^2 \\ d &= 39\ \text{m} \end{aligned}$$

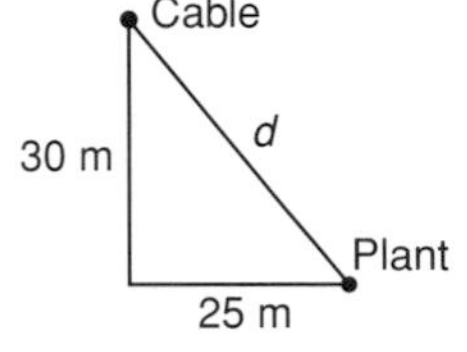

Example 2

An escalator rises 15.0 m in a distance of 18.0 m. How long is the escalator?

$\text{Length}^2 = 15^2 + 18^2 = 225 + 324 = 549\ \text{m}^2$

$\text{Length} = \sqrt{549} = 23.4\ \text{m}$

Question to try

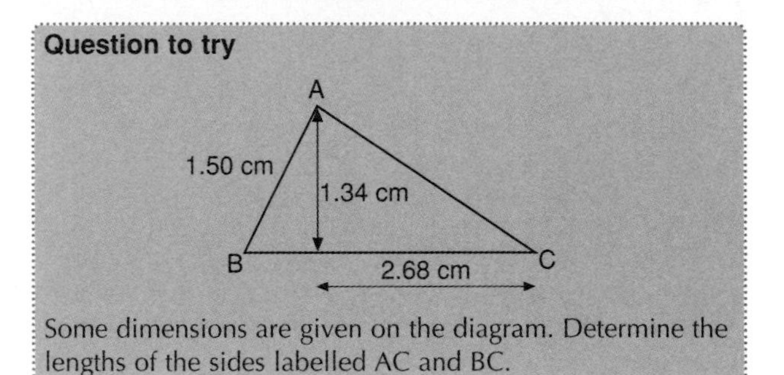

Some dimensions are given on the diagram. Determine the lengths of the sides labelled AC and BC.

Finding angles

It is sometimes necessary to determine an angle in a diagram when you know another. This is often necessary when working with forces and in ray diagrams in optical work. The properties of triangles are useful when doing this and there are a few other useful things that help.

Look for the Z shape: the angles marked are equal.

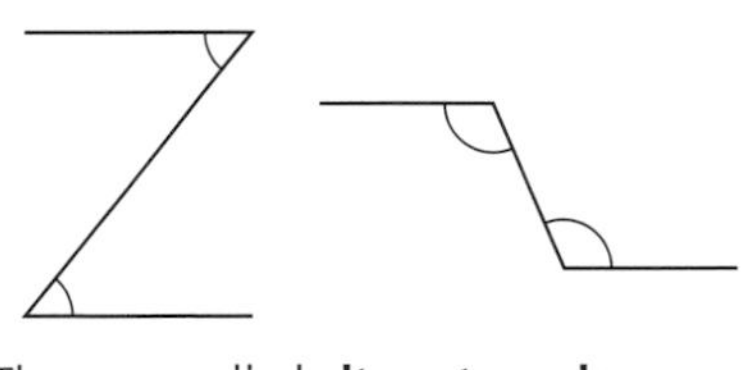

These are called **alternate angles**.

The angles marked are equal.

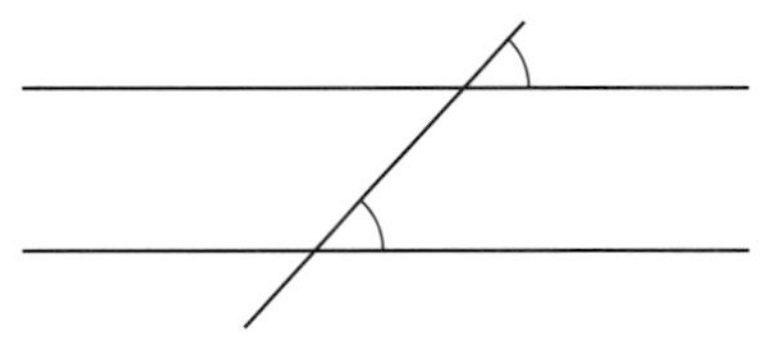

These are called **corresponding angles**.

The angles marked are equal.

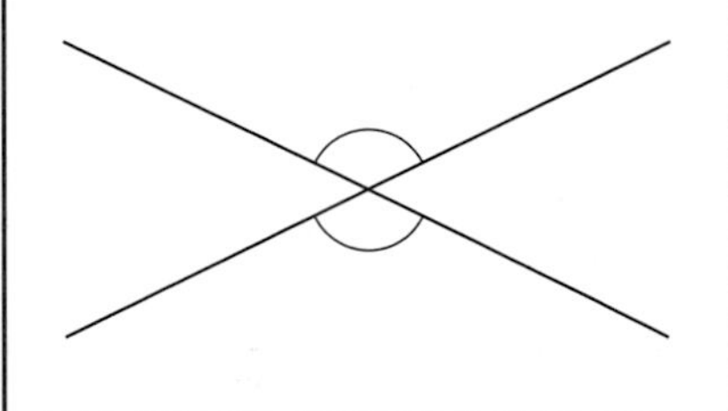

The angles marked add up to 180°.

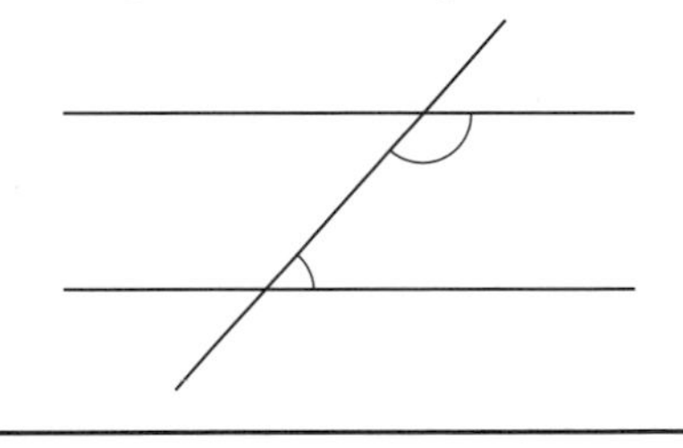

The triangle is a right-angled triangle. The angles marked are equal.

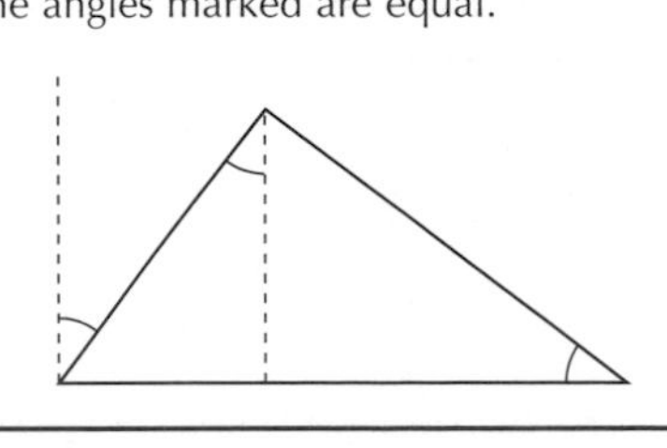

Questions to try

In each diagram determine as many angles as you can

1.

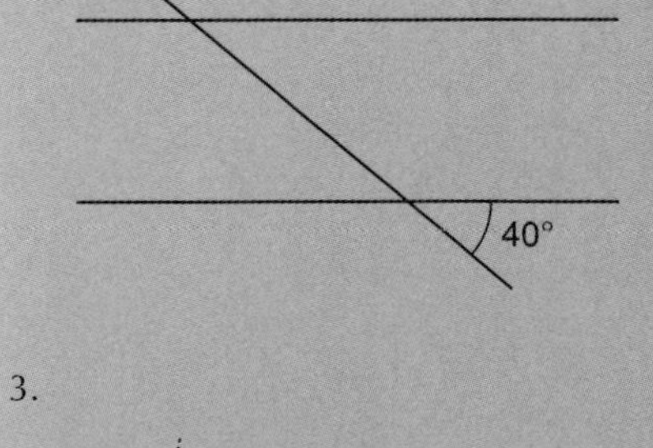

2.

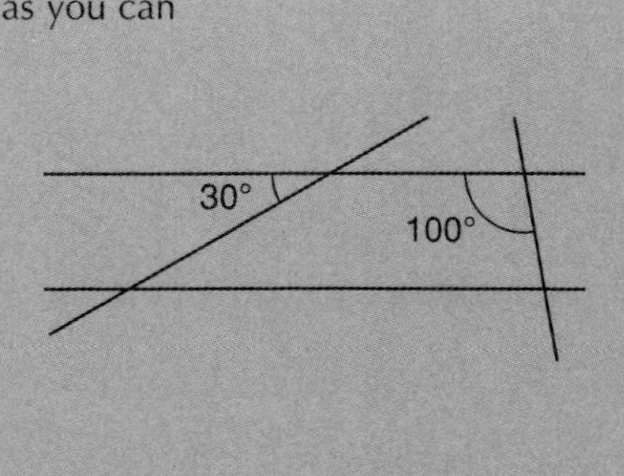

3.

90°
90°
10°

4.

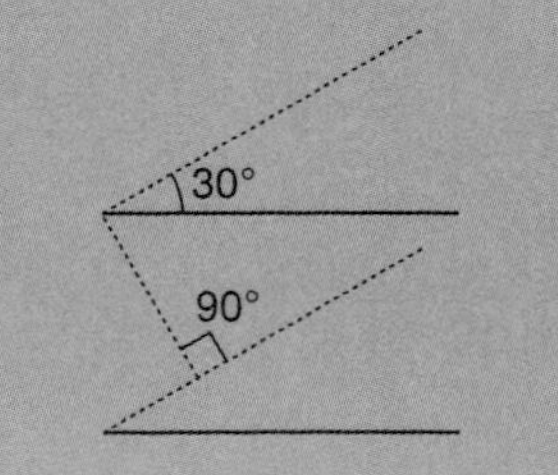

Useful facts about sines, cosines, and tangents

This page contains some useful relationships between sines and cosines and some commonly used values. Knowing these can simplify many problems.

In the diagram

$\sin \theta = \cos \varphi$, so

$\sin \theta = \cos (90 - \theta)$

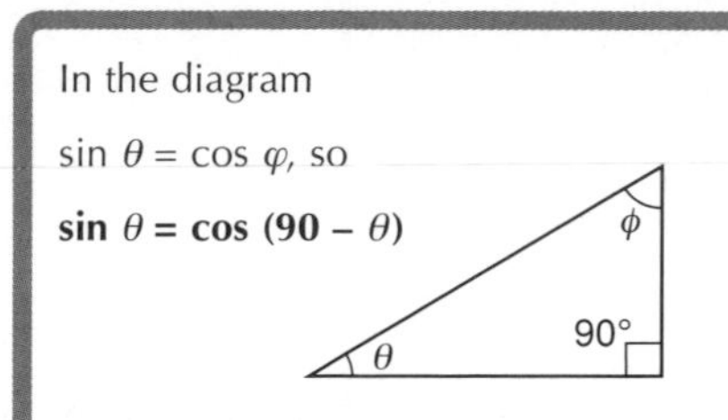

$\sin 45 = \frac{1}{\sqrt{2}}$

$\cos 45 = \frac{1}{\sqrt{2}}$

$\tan 45 = 1$

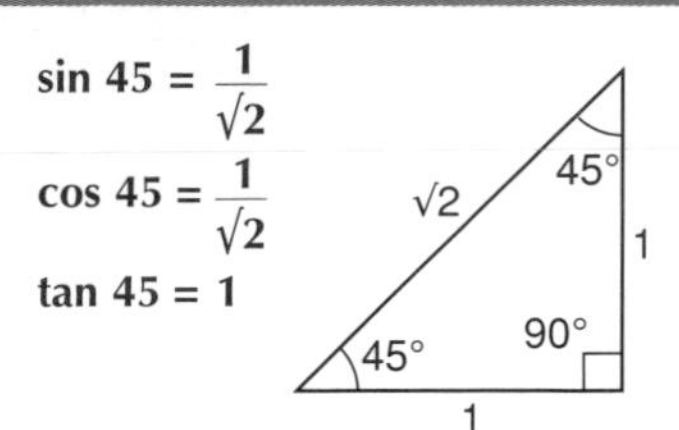

$\sin 30 = \cos 60 = \frac{1}{2}$

$\cos 30 = \sin 60 = \sqrt{\frac{3}{2}}$

$\tan 30 = \frac{1}{\sqrt{3}}$

$\tan 60 = \sqrt{3}$

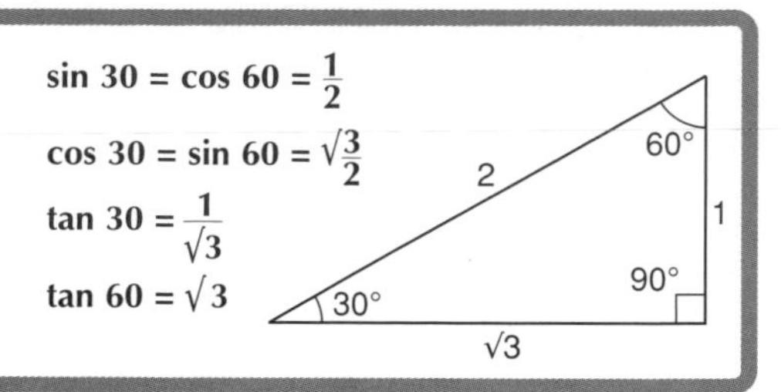

Combining the definitions of sine and cosine and using Pythagoras' theorem given on page 54, we can show that:

$\sin \theta = \cos (90 - \theta)$

$\sin^2\theta + \cos^2\theta = 1$

Small angle approximations

For small angles (less than 10°)

$\sin \theta \approx \tan \theta \approx \theta$ **in radians**

$\cos \theta \approx 1$

For example:

sin 3 = 0.0523
tan 3 = 0.0524
θ in radians = (3/360 × 2π) = 0.523
cos 7 = 0.99

Resolving vectors

Sines and cosines are used extensively in work with vector quantities.

Distinction between a vector and a scalar quantity

- A **vector** quantity has **magnitude** and **direction**.
- A **scalar** quantity has **only magnitude**.

Note: Examples of vector quantities

Force, weight, displacement, momentum, electric and gravitational field strengths.

Note: Examples of scalar quantities

Speed, energy, mass, amount of substance, number of particles.

Representation of vector quantities

A vector quantity is represented by a line with an arrow.

- The direction the arrow points is the direction of the vector quantity.
- The length of the line represents the magnitude (size) of the vector quantity, e.g. a line of length 5.0 cm could represent a force of 10 N using a scale of 2 N = 1 cm.

Any vector quantity can be resolved into two components at right angles to each other

The directions are often the x and y directions, but any two directions at right angles could be used.

The **effective value** of a vector in any direction can be found by resolving in that direction.

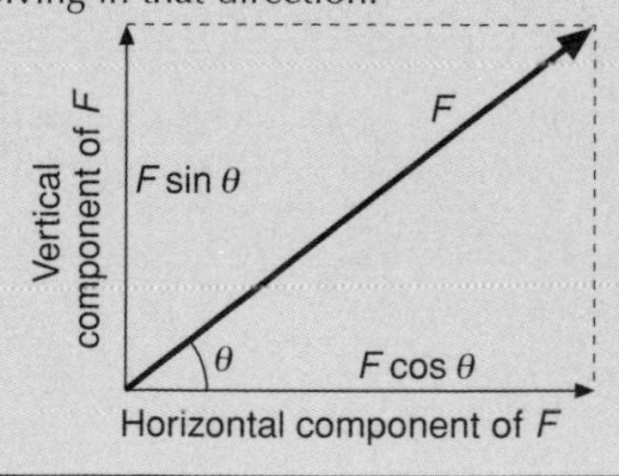

Example 1: Resolving forces on a slope

A body of weight W rests on a slope of angle θ. Write down an equation for the effective force down the slope. What is the normal reaction to the surface?

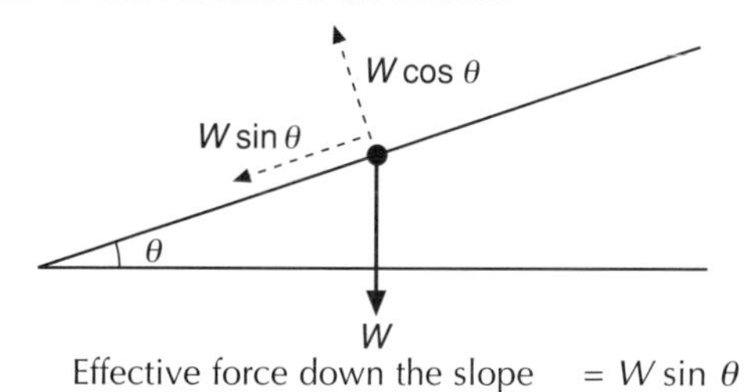

Effective force down the slope $= W \sin \theta$
Normal reaction $= W \cos \theta$

Example 2

An aircraft coming in to land has a velocity of 200 km h^{-1} at an angle of 5.0° to the horizontal. Determine the vertical speed of the aircraft.

Vertical speed = 200 sin 5 = 17.4 km h^{-1}

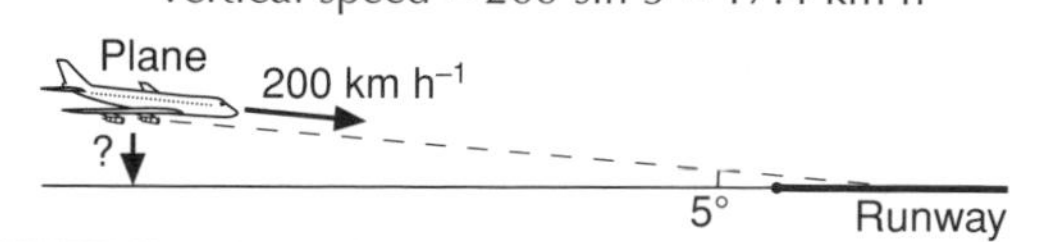

Example 3

In the British Isles the Earth's magnetic field strength is 4.4×10^{-5} T (tesla) at an angle of 66° to the horizontal. Only the horizontal magnetic field affects an orienteering compass. What is the magnitude of the horizontal component of the Earth's field in the British Isles?

Horizontal component $= 4.4 \times 10^{-5} \cos 66$ T
$= 1.8 \times 10^{-5}$ T

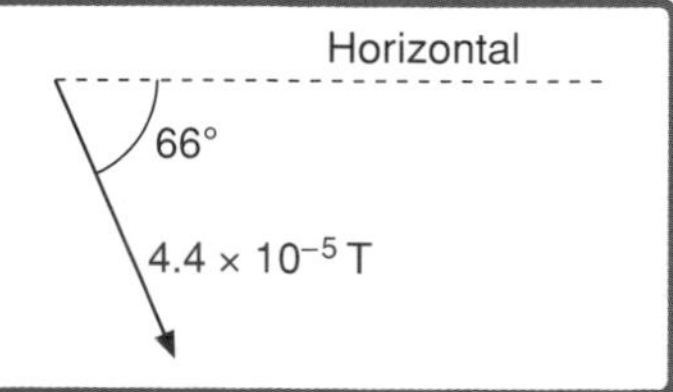

Sine and cosine curves

There are many instances in scientific work when a quantity is found to vary sinusoidally with position or time. It is important to understand what this means and how to check whether experimental data vary sinusoidally.

What is a sinusoidal variation?

A sinusoidal curve is a graph of the sine of an angle against the angle itself.

The angle may be given in degrees or radians. Angles in radians are commonly used.

Sinusoidal variations with time

Some quantities vary sinusoidally with time, for example:

- the displacement of a mass oscillating on the end of a spring
- the voltage of the a.c. mains supply.

$+A$, 0, $-A$, A, $\frac{1}{4}T$, $\frac{1}{2}T$, $\frac{3}{4}T$, T, $\frac{5}{4}T$, $\frac{3}{2}T$

T is the period of the wave (this corresponds to 360°)

You can only take the sine of an angle, so how can you have sinusoidal variations with time?

When a quantity x varies sinusoidally, the equation that shows how x varies with angle θ is

$$x = A \sin \theta$$

In many practical situations θ changes with time at a constant rate (the angle term is directly proportional to time) and the equation becomes

$$x = A \sin \omega t$$

ω is the **angular frequency**. This determines how long it takes for a complete cycle of angle changes to occur. The time for one complete cycle is the **period,** T.

$\omega = 2\pi f$, where f is the **frequency**, i.e. the number of cycles of the sine wave competed each second.

For a pendulum f is the number of oscillations per second; for an alternator it is the number of revolutions of the coil per second.

Difference between a sine curve and a cosine curve

These have the same basic shape.

The cosine curve has a maximum value when the angle is 0° whereas the sine is zero at 0°. At 90° the reverse is true.

We say that the sine and cosine curves are **90° out of phase**. Because the cosine curve reaches the maximum first it is said to be **leading** the sine curve by 90° (or $\pi/2$ radians).

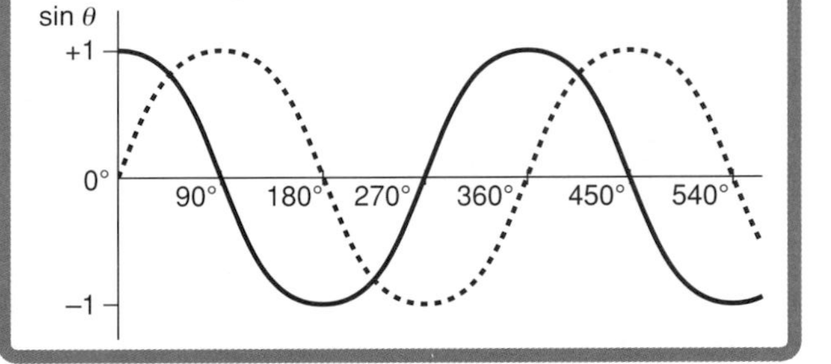

Checking whether a curve is sinusoidal

A graph may look sinusoidal but is it really?

To check, perform the following steps:

- Measure the amplitude A.
- Calculate the displacement at a number of angles such as 30°, 45° and 60° using the formula $A \sin \theta$.
- Imagine the axis to be in terms of angle rather than time, remembering that one complete cycle is equivalent to 360°. (This can be done because the angle is a constant multiplied by t.)
- Plot the calculated values on the curve being checked.
- If they agree within the limits of uncertainties in the practical curve, the curve is sinusoidal.

Example

Check whether this curve is a cosine curve.

- The maximum value is 75 mm.
- At $t = 1$ s the angle term is $\pi/4$.
- The value should be 75 cos 45 = 53 mm.
- When this is plotted on the curve it does not match the practical value.
- Do this for other points to check.
- None of the points match except at the peak and when the displacement is zero so the curve is *not* a cosine curve.

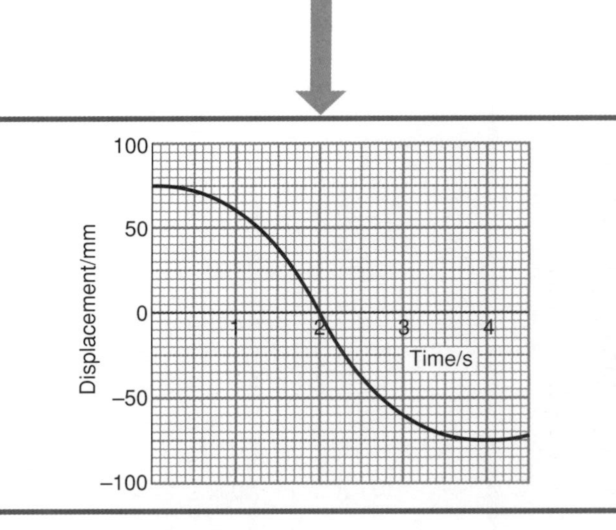

Note: Checking other experimental data

Plotting theoretical data obtained from a possible formula on a graph together with experimental data is a good way of checking whether theory and practice agree. The principle can be used to compare any theoretical and experimental data, not only sinusoidal variations.

The wave equation

Waves can be traced back to an oscillation. If the oscillator producing the wave has a displacement that varies sinusoidally with time, the wave produced along the medium has a sinusoidal profile. The wave equation describes both parts of this motion, the time dependence and the dependence on distance along the medium carrying the wave.

Producing a sine wave

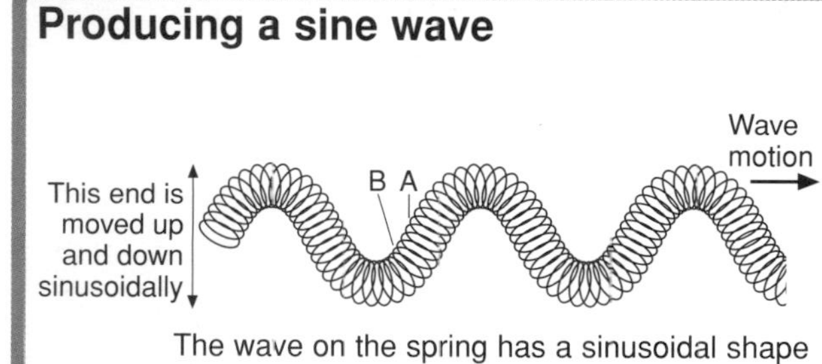

The wave on the spring has a sinusoidal shape

Graph of displacement against time

The displacement of each coil in the spring varies sinusoidally with time. Each turn in the spring oscillates identically but slightly **out of phase** with the coil next to it.

Turn A perfoms the same movements as turn B but a little later.

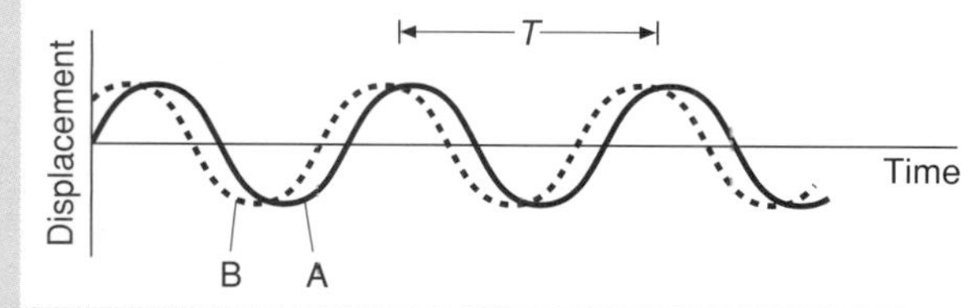

Each turn of the spring repeats the cycle of changes in a time T (called the period of the oscillation).

Displacement–time equation for a particle in the medium

The equation representing the variation of displacement of each particle with time is

$$y = A \sin\left(\frac{t}{T}2\pi\right)$$

In this equation y is the displacement and A is the maximum displacement.

Whenever the time t changes by the period T, the sine changes by 2π. This means that the displacement is back where it started every T seconds.

Variation of displacement with distance along the spring

This is a 'snapshot' of the wave on the spring taken at a particular time. A snapshot taken a little later would show that the wave had moved to the right.

Snapshot a little later
Displacement
Distance along spring
Snapshot at time t
λ

This shows that at a given time particles have different displacements but the wave repeats itself at certain distances along the medium. This is called the wavelength, λ.

Displacement–position equation for a particle in the medium

The equation representing the variation of each particle with distance x along the medium is

$$y = A \sin\left(\frac{x}{\lambda}2\pi\right)$$

Whenever the distance x changes by λ the sine changes by 2π and the displacement is back where it started.

The wave equation

The equations for the time variation and the variation with distance from the source can be combined to enable the displacement of any particle in the medium at any time to be determined.

$$y = A \sin\left(\frac{t}{T}2\pi + \frac{x}{\lambda}2\pi\right)$$

$$y = A \sin 2\pi\left(\frac{t}{T} + \frac{x}{\lambda}\right)$$

This equation assumes that when timing begins ($t = 0$) the displacement at the source is zero.

Try a few calculations at different times and distances along the medium to practise using the equation.

Example

In a wave each particle has an amplitude $A = 0.150$ m and a period of oscillation $T = 2.00$ s.

The displacement–time equation gives the following values for the displacement y at times t.

***t* / s**	0	0.25	0.50	0.75	1.00
***y* / m**	0	0.106	0.150	0.106	0

***t* / s**	1.25	1.50	1.75	2.00	2.25
***y* / m**	–0.106	–0.150	–0.106	0	0.106

The calculation at 0.25 s gives the displacement as

$$y = 0.150 \sin\left(\frac{0.25}{2.0}2\pi\right)$$

$$= 0.150 \sin\left(\frac{\pi}{4}\right)$$

$$= 0.150 \sin 45 = 0.106$$

You may wish to check other values for yourself. The angle in the formula is in radians (2π radians = 360°).

This shows that the equation correctly predicts the cycle to be repeated after a time equal to the period.

Question to try

A wave has amplitude 0.150 m. The wavelength is 0.400 m. Calculate the displacements of the particles at the following distances along the medium when the displacement at the source is zero.

***x* / m**	0	0.050	0.100	0.150	0.175
***y* / m**					

***x* / m**	0.200	0.250	0.300	0.400	0.450
***y* / m**					

The calculations are similar to those for the time variation. They show that the equation correctly predicts the cycle to be repeated after a distance equal to the wavelength.

Root mean square voltages

When working with alternating voltages it is necessary to know the root mean square (r.m.s) voltage and the r.m.s. current. These are the voltages and currents that enable us to determine how much power we are using. Root mean square values apply when the voltage is varying sinusoidally, as is the case for the mains supply.

Mean value of voltage over a cycle

The equation that tells us that the instantaneous voltage v at time t is $v = V_{pk} \sin(2\pi ft)$.

- v is the instantaneous voltage.
- V_{pk} is the maximum positive or negative voltage (the peak voltage).
- f is the frequency of the supply.

In a cycle of a sinusoidal variation $\sin \theta$ is positive and negative equally, so the sum of all the values and hence the average is 0.

The mean value of the voltage over a complete cycle is zero.

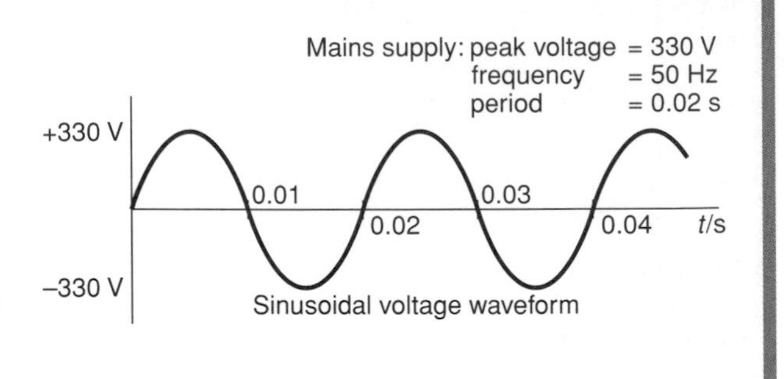

Sinusoidal voltage waveform

Why is the root mean square voltage useful?

The alternating voltage produces heat in a resistor, so using the mean voltage does not tell us anything about how much power on average is dissipated in the resistor.

Power = V^2/R so the power that is being generated at a time t is

$$\frac{[V_{pk} \sin(2\pi ft)]^2}{R} = \frac{V_{pk}^2}{R} \sin^2(2\pi ft)$$

The peak power V_{pk}^2/R is dissipated when $\sin(2\pi ft) = 1$. ($\sin^2(2\pi ft)$ is then also 1.)

The mean power over a cycle (which tells us how much power is generated on average) is the important quantity.

Mean power = mean value of $\sin^2(2\pi ft)$ × peak power

Since the mean of the $\sin^2$ over a cycle $= \frac{1}{2}$,

$$\text{Mean power dissipated} = \frac{V_{pk}^2}{2R}$$

$$\text{Mean power} = \frac{V_{rms}^2}{R} \quad \text{where } V_{rms} = \frac{V_{rms}^2}{\sqrt{2}}$$

This shows that when calculating the power dissipated in a circuit it is the r.m.s. voltage that is important and not the mean voltage.

The mean power dissipated in the resistor by the alternating voltage is the same as that dissipated in the same resistor by a direct voltage of magnitude $V_{rms}/\sqrt{2}$.

If you want to know more about:
Sines and cosines see pages 53, 55

A $\sin^2\theta$ curve

The sine squared curve is obtained from the sine curve by squaring each value. Since both positive and negative values give a positive value when squared, there are no negative values in the $\sin^2\theta$ graph.

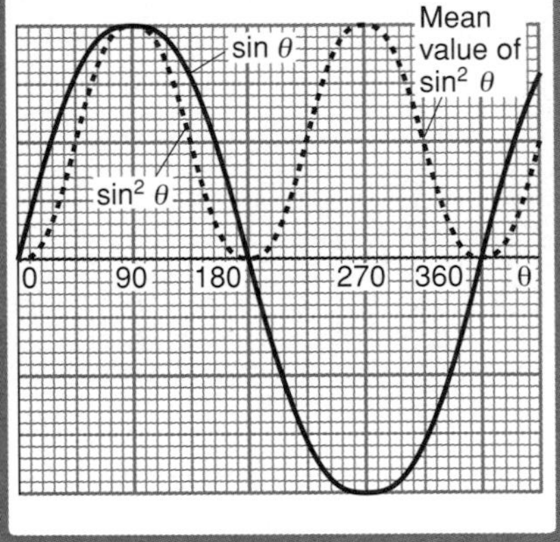

Mean of $\sin^2\theta$ over a cycle

The $\sin^2\theta$ graph is symmetrical about half the peak value. The mean of all the values of $\sin^2\theta$ over one cycle is therefore $\frac{1}{2}$ because the maximum value is 1^2, which is also 1.

The mean value of $V_{pk}^2 \sin^2(2\pi ft)$ is $V_{pk}^2/2$. This is the mean square voltage of the supply.

The square root of this, $V_{pk}/\sqrt{2}$, is the root mean square or r.m.s. voltage of the supply.

Other voltage waveforms?

Signal generators have one output that varies sinusoidally and other outputs that vary in a form called a square wave or a triangular wave.

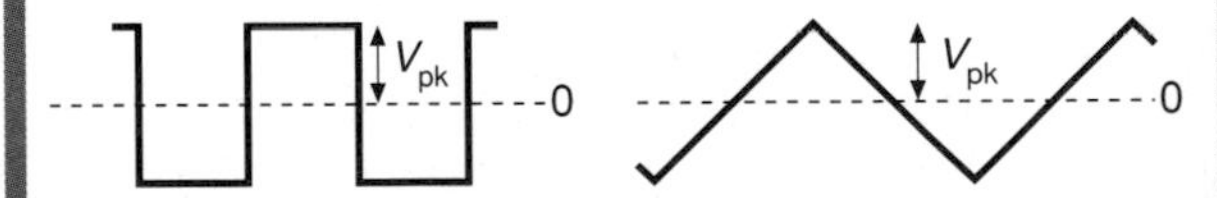

The waveforms here are alternating but the r.m.s. voltage for these is **not** $V_{pk}/\sqrt{2}$.

For a square waveform that has equal positive and negative peak voltages, the voltage to use when calculating the average power generated is the peak voltage.

For a triangular wave with equal positive and negative peak voltages the r.m.s. voltage is $V_{pk}/\sqrt{3}$.

Questions to try

1. Calculate the r.m.s. voltage when the peak voltage is 120 V.
2. (a) Calculate the positive peak voltage of the 230 V mains supply.
   (b) What is the voltage difference between the positive and negative peak values?

Root mean square speeds

Why is the root mean square speed important?

In the kinetic theory of gases it is found that the temperature of a gas is related to the mean kinetic energy (KE) of the gas molecules.

Mean kinetic energy of a molecule $= \frac{1}{2} m <v^2> = \frac{3}{2} kT$

where m is the mass of a molecule, $<v^2>$ (or $\overline{v^2}$) is the mean square speed, T is the temperature and k is the Boltzmann constant.

This shows that the temperature is proportional to the mean KE of the gas. The mean KE is proportional to the mean square speed of the molecules.

Calculation of mean square speed

The mean square speed of the molecules is related to the energy of the gas, so this is more relevant than the mean speed.

The **r.m.s. speed** is calculated as follows:

1. Square the speeds of each individual molecule.
2. Add these together.
3. Divide by the number of molecules.
4. Take the square root of the answer.

Mathematically this would be written as

$$v_{rms} = \sqrt{\left(\frac{\sum_1^n v_n^2}{n}\right)}$$

This means that all the values of v^2 (v_1^2, v_2^2...up to v_n^2) are added together and divided by the number of molecules. The square root of this is then taken to arrive at the r.m.s. value.

Example

Six molecules have speeds in m s^{-1} of 300, 400, 500, 600, 700 and 800. Find (a) the r.m.s. speed and (b) the mean speed.

(a) Mean square speed

$$= \frac{300^2 + 400^2 + 500^2 + 600^2 + 700^2 + 800^2}{6}$$

$= 331\ 700$ m^2 s^{-2}

The r.m.s. speed is 576 m s^{-1}.

(b) Mean speed

$$= \frac{300 + 400 + 500 + 600 + 700 + 800}{6}$$

$= 550$ m s^{-1}

This gives a mean speed squared of 302 500 m^2 s^{-2}, which is quite different from the mean square speed calculated above.

The Maxwell distribution

The speeds of molecules in a gas follow a distribution called the Maxwell distribution.

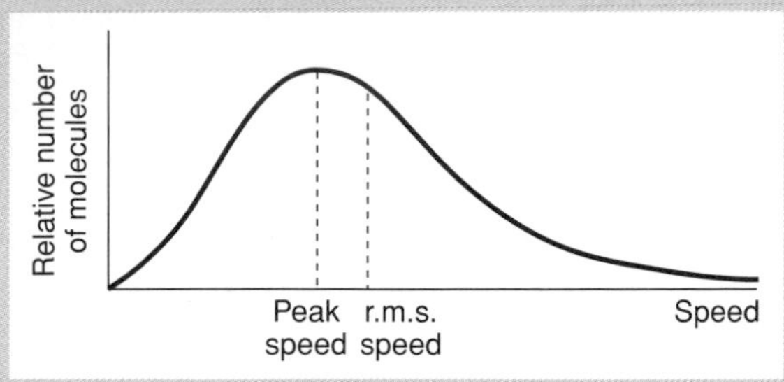

The distribution is not symmetrical. The peak shows the most probable speed.

The range of molecular speeds is from zero to a high value, but very few molecules have a speed more than four times the r.m.s. speed.

The mean kinetic energy of molecules of different gases at any given temperature is the same, so that molecules of low mass can have very high speeds. Even at normal Earth temperatures of around 300 K some will have speeds in excess of the speed necessary to escape from the Earth and will leave the atmosphere. This explains why there is a low concentration of gases such as helium in the atmosphere.

Note: $\overline{v^2}$ is an alternative way of writing $<v^2>$.

Note: the r.m.s. speed is **not** the same as the mean speed. This would be found by adding all the speeds together and dividing by the number of molecules

$$\text{Mean speed} = \frac{\sum_1^n v_n}{n}$$

Question to try

The bar chart shows the distribution of speeds of a number of molecules. Determine the r.m.s. speed of the molecules.

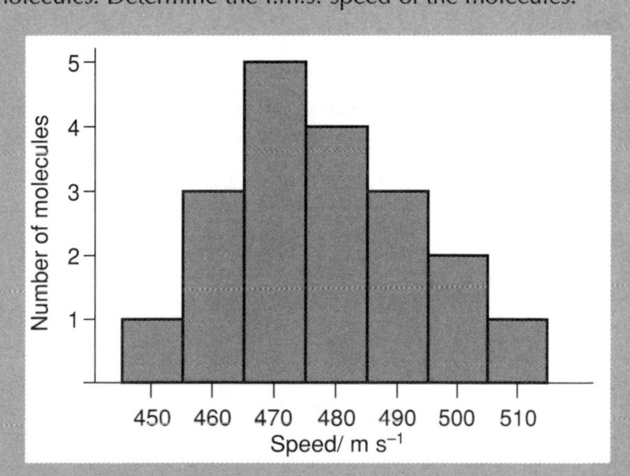

If you want to know more about:

Frequency distributions see page 71

Mean values see page 73

Adding and subtracting vectors using scale diagrams

Scalar quantities are added and subtracted using simple arithmetic. When adding and subtracting vector quantities account has to be taken of their direction. This can be done either by scale drawings or by using the properties of triangles (i.e. by trigonometry).

If you want to know more about:

Standard form see pages 10–11
Scaling see pages 32–33

Resultant

The result of adding or subtracting two vectors is called the **resultant**. It is the single vector that has the same effect as the two vectors.

Adding two vector quantities P and Q acting in different directions

This can be found by drawing a **scale diagram** and using a **vector parallelogram**.

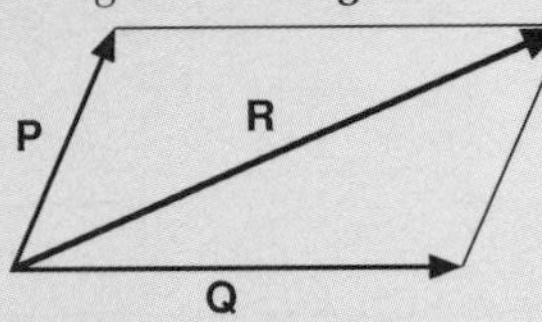

1. Choose a scale so that the lengths of the lines representing the sizes of **P** and **Q** can be drawn in the available space.
2. Draw the lines so that their directions are in the direction in which **P** and **Q** act. These are two adjacent sides of the vector parallelogram.
3. Complete the parallelogram.
4. Draw the diagonal that lies between **P** and **Q**.
5. The length and direction of the diagonal **R** represents the size (magnitude) and direction of the sum of **P** and **Q**.

Example: A trolley on a ramp

A trolley of weight 1.2 kN is pulled up a smooth ramp that makes an angle of 25° to the horizontal by a steady force of 2.4 kN. Determine the resultant force acting on the trolley.

In the scale used here, 1.0 cm represents 0.50 kN. In practice a much larger scale is needed to improve accuracy. (You may like to try this using a larger scale.)

The resultant force is 2.2 kN at 85° to the downward vertical.

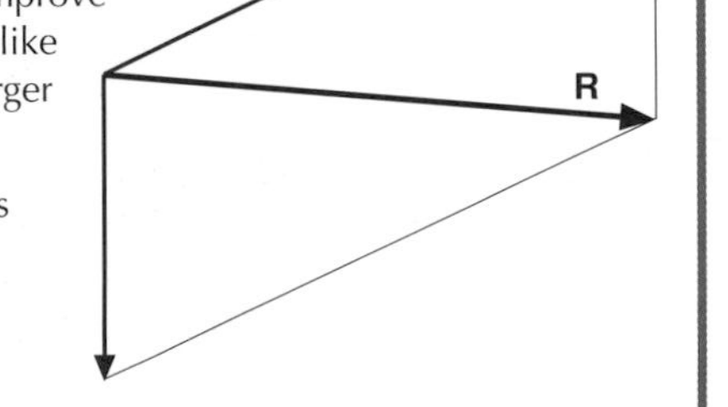

Resultant of two vector quantities acting in the same or opposite directions

When the directions are the same or opposite, the magnitudes can be added using simple arithmetic.

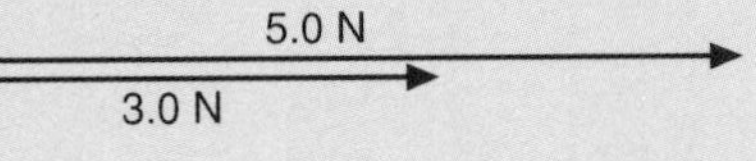

The resultant force of the two forces is 8.0 N to the right.

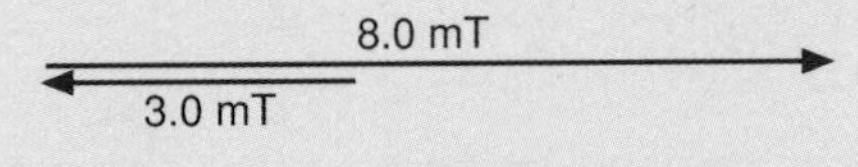

The resultant of the two magnetic field strengths is 5.0 mT to the right.

Questions to try

1. Two gravitational fields of magnitude 0.075 N kg^{-1} and 0.030 N kg^{-1} act in the directions shown. Determine the resultant gravitational field.

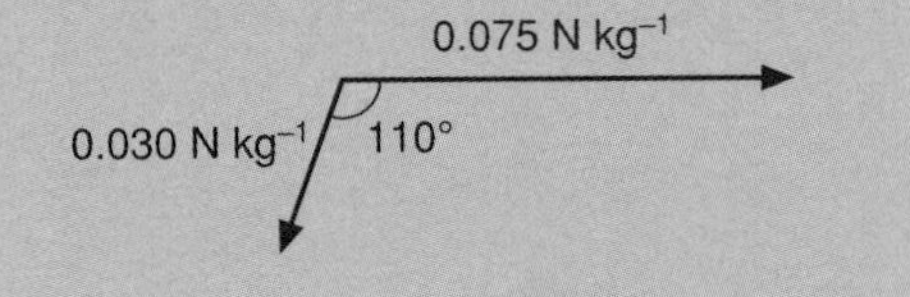

2. A car moves at 20 m s^{-1} to the east. Determine the velocity that has to be added to it so that it is moving at a speed of 30 m s^{-1} to the north-east.
3. Determine the difference between a momentum of 40 N s to the south and a momentum of 50 N s to the north-east.

Finding the difference between two vectors P and Q

As with addition, the difference between the vectors, **Q** − **P**, can be found by drawing a **scale diagram**.

1. Choose a scale so that the lengths of the lines representing **P** and **Q** can fit the available space.
2. Draw the lines to scale and so that their directions are those in which **P** and **Q** act.
3. Draw the line that joins the ends of the vectors together.
4. The length of this line represents the magnitude of the difference between the two vectors.
5. The direction of the vector **Q** − **P** is in the direction from the end of **P** to the end of **Q**.

Example: Electron motion

An electron in a television tube changes direction when passing between the deflecting coils. It enters at a speed of 1.5×10^7 m s^{-1} horizontally and leaves at the same speed travelling at 20° to the horizontal in the upward direction. Determine the difference between the final and the initial velocity.

The scale used here is 2 cm to represent 0.5×10^7 m s^{-1}.

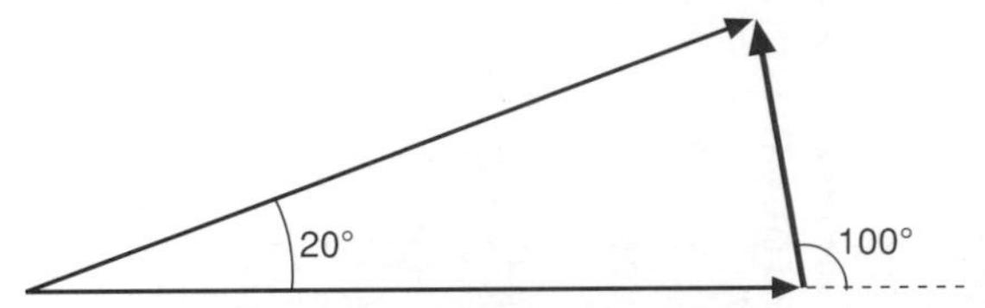

The difference between the initial and final velocities is 1.0 m s^{-1} in a direction at 100° to the original direction of the motion. Looking at it another way, to reach the final velocity a velocity of 1.0 m s^{-1} in a direction 100° to the original direction has to be added to the initial velocity.

Adding and subtracting vectors by calculation

There are mathematical rules that enable you to add or subtract vectors at any angle to each other. However, the important case that you will use most often is when they are at right angles to each other. Other resultant vectors can be found by first resolving into two vectors at right angles.

Adding vector quantities at right angles

The properties of right-angled triangles and Pythagoras's theorem can be used.

The process is easiest to show assuming that the two vector quantities are vertical v and horizontal h but the process is the same for any vector quantities at right angles.

The vector parallelogram in this case is a rectangle.

1. The size of the resultant is $\sqrt{(v^2 + h^2)}$.
2. The direction is at an angle to the horizontal given by $\tan\theta = v/h$, so $\theta = \text{INV} \tan (v/h)$ or $\tan^{-1} (v/h)$.

Example

A case of medical supplies is dropped without a parachute from an aircraft. As it falls its horizontal momentum is constant at 7500 kg m s^{-1}. It hits the ground with a vertical momentum of 5400 kg m s^{-1}. Calculate the total momentum on impact.

7500 kg m s^{-1}
5400 kg m s^{-1}
Total momentum **R**

R is the total momentum. It is the vector sum of the horizontal and vertical momentum.

To find the magnitude use Pythagoras's theorem:

$$\mathbf{R}^2 = (7500^2 + 5400^2) \text{ kg m s}^{-1}$$
$$= 8.5 \times 10^7 \text{ kg m s}^{-1}$$
$$\mathbf{R} = 9200 \text{ kg m s}^{-1}$$

To find the direction use the tangent rule:

$$\tan\theta = o/a = 5400/7500 = 0.72$$

The angle of the momentum to the horizontal on landing is therefore $35.8° \approx 36°$.

Resultant of any two vector quantities

1. Resolve each of the vectors in two directions at right angles, usually horizontally and vertically.
2. Add the horizontal and vertical components.
3. Proceed as for two vector quantities at right angles.

It is likely that you will be working with only two or three vector quantities, but the principle is the same for any number.

Example

Two electric fields, one of strength 200 V m^{-1} and the other of strength 150 V m^{-1}, act at 60° to one another as shown in the diagram. Calculate the resultant field strength.

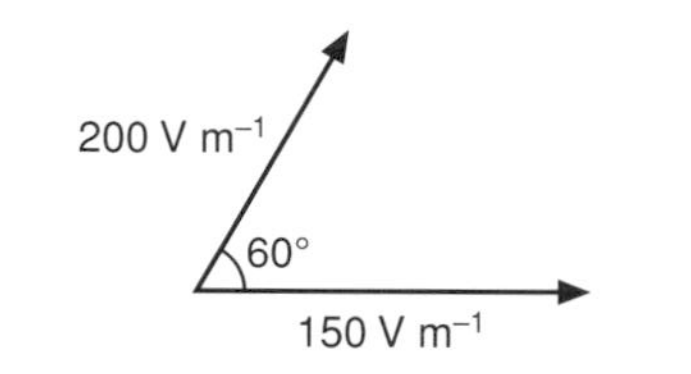

Vertical component of the 200 V m^{-1} field = 200 sin 60 V m^{-1}
Horizontal component of the 200 V m^{-1} field
= 200 cos 60 V m^{-1}

Total vertical component = 173 V m^{-1}
Total horizontal component = 150 + 200 cos 60 = 350 V m^{-1}

Resultant field strength = $\sqrt{(173^2 + 350^2)}$ = 390 V m^{-1}

Angle to horizontal (the 150 V m^{-1} field) = INV tan (173/350)
= 26°

Note: Draw a diagram
Even when a calculation is to be used it is wise to have a diagram of the vectors to work with. This need only be a sketch but should contain all the known information.

Difference between two vector quantities

This process will be required less often than addition in your course. An important case, and the only one we will deal with here, is the case of two vectors of equal size F with an angle θ between them.

- The difference between them is $2F \sin(\theta/2)$.
- The direction of the change (the difference) is $90 + \theta/2$ to the initial vector.

Example

The direction of motion of a particle moving in a circular path at a constant speed of 3.5 m s^{-1} changes by an angle of 10°. What is the change in the velocity?

Change = 2 × 3.5 × sin (10/2) = 0.61 m s^{-1}

The direction of the change is at an angle of 90 + (10/2) = 95° to the original direction in which the particle was going.

Questions to try

1. At one place on the Earth's surface the horizontal component of the Earth's magnetic field is 1.8×10^{-5} T. The vertical component is 3.1×10^{-5} T. Calculate the magnitude and direction of the resultant magnetic field strength.
2. A mass is acted on by a force of 15 N. Calculate the force that has to be applied at right angles to this force so that the resultant is a force at an angle of 30° to it.
3. Calculate the velocity change of a particle moving at 1.5×10^6 m s^{-1} when it turns through an angle of 5° without changing speed.

If you want to know more about:
Sines — see pages 53, 55
Pythagoras's theorem — see page 54

Degrees and radians

These are both units in which angles may be measured. The radian is used in many instances rather than the degree.

Degrees

A degree is 1/360 of the angle that a radius arm (such as a single spoke on a bicycle wheel) sweeps out when it completes one revolution of a circle.

We say that the angle subtended at the centre of a circle by the circumference is 360°.

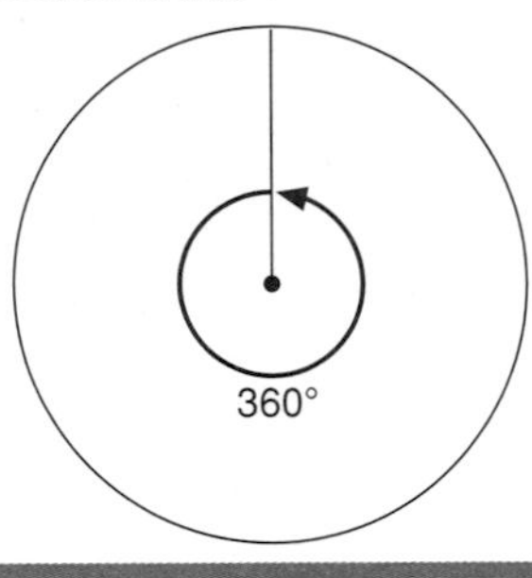

To convert angles from degrees to radians

$$\text{Angle in radians} = \frac{\text{angle in degrees}}{360} \times 2\pi$$

To convert angles from radians to degrees

$$\text{Angle in degrees} = \frac{\text{angle in radians}}{2\pi} \times 360$$

Radians

A radian (abbreviated rad) is the angle subtended at the centre of a circle by an arc that has the same length as the radius of the circle.

$$\text{Angle in radians} = \frac{\text{arc length}}{\text{radius}}$$

A complete circle has a circumference of $2\pi r$. The angle subtended at the centre of the circle by the circumference is therefore

$$\frac{2\pi r}{r} = 2\pi \text{ rad}$$

It follows that

2π rad is the same as 360°
π rad is the same as 180°
$\pi/2$ rad is the same as 90° etc.

Note: Do not forget to include a value for π

It is not uncommon to put angles in terms of π in working. Mathematicians often give answers in terms of π; however, this should not be done in scientific work. The final answer must be worked out and quoted to a number of significant figures that is consistent with the data.

Small angle approximations

It is sometimes possible to produce a simpler mathematical relationship between two quantities by restricting the relationship to small angles. Particular cases where this is used are the theory of the simple pendulum and the production of interference fringes in Young's two-slit experiment.

When angles are larger you cannot simplify the equation and the theory becomes more complicated.

The approximations made depend on the fact that, provided the angles are small enough, to a good approximation

$\sin \theta = \tan \theta = \theta$ in radians

This is easy to see in a diagram:

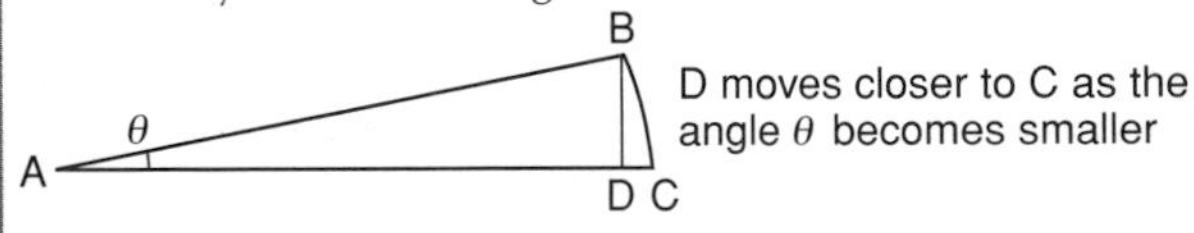

$\sin \theta = BD/AB$ and $\tan \theta = BD/AD$

When the angle becomes small $AB \approx AD$ so $\sin \theta \approx \tan \theta$.

Angle in radians = arc BC/AB

When the angle is small $BC \approx BD$ so the angle in radians $\approx \sin \theta$.

For an angle of 10° the values of the sine, tangent and angle in radians differ by about 1%, and the difference is less for smaller angles. The effect of making an approximation would therefore be noticeable only in accurate experimental work. You can verify this by doing question 3.

Note: What happens to cosines when angles are small?

This is used less often but it is sometimes useful to note that, when the angle is small,

$\cos \theta \approx 1$

Questions to try

1. Convert the following to radians:

   (a) 45° (b) 5° (c) 75°
   (d) 270° (e) 450° (f) 2.5×10^{4}°

2. Convert the following angles in radian to degrees:

   (a) 2.0 (b) 3π (c) 0.50
   (d) 300 (e) 2000π

3. Check using your calculator how close the sine, tangent and angle in radians are for the following angles:

   1° 5° 10° 12° 15°

   (You should find that for angles up to 10° the difference is within 1 or 2 in the third significant figure. They are the same to about 1%. The difference is much larger for the larger angles.)

4. (a) What are the cosines of the angles in question 3?

   (b) For what angle would an approximation of $\cos \theta = 1$ give a 1% error in the calculation?

Non-linear motion

We know that objects can move in curved paths as well as in straight lines. A stone thrown from a cliff, an electron between the deflecting plates of a cathode ray oscilloscope, particles in accelerators and cars on roundabouts all move in curved paths. Some paths can be quite complex but you need only be familiar with two relatively simple paths in your studies. These are circular motion and parabolic paths.

UNIFORM MOTION IN A CIRCLE

In translational motion we describe the motion of an object by referring to its displacement, velocity and acceleration. A number of similar angular terms are used to describe rotational motion.

Although the **linear speed** v is constant, the **linear velocity** is changing. This is because the direction of motion is changing continuously.

The particle is accelerating. This acceleration is toward the centre of the circular path and is called the **centripetal acceleration**. The magnitude of the centripetal acceleration is

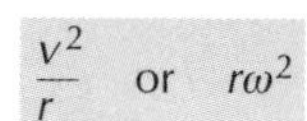

$\frac{v^2}{r}$ or $r\omega^2$

The unit is **m s^{-2}**.

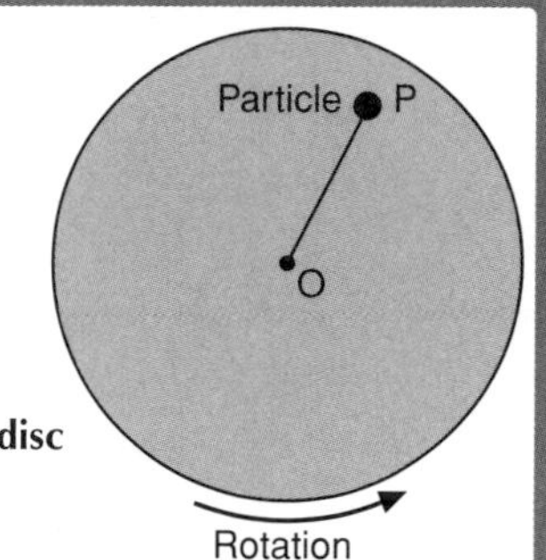

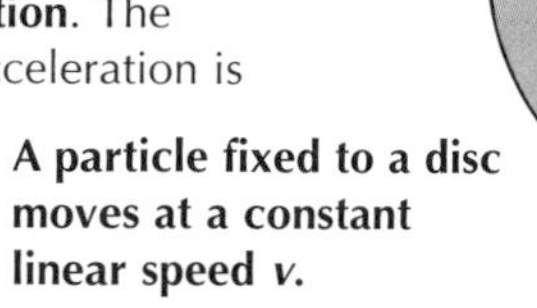

A particle fixed to a disc moves at a constant linear speed *v*.

Angular displacement θ

This is the angle swept out in relation to a reference position by the radial line OP.

It is usual to measure angles in radians when working with circular motion.

Angular velocity ω

This is the angle swept out per second by the radial line OP.

The angle θ is swept out in a time t

$$\omega = \frac{\theta}{t}$$

The unit is **rad s^{-1}**.

Relationships between linear and angular velocity

$$v = r\omega$$

The particle P moved through an arc of length v in 1s.

The angle in radians swept out by OP = ω:

$$\omega = \frac{\text{arc length}}{\text{radius}} = \frac{v}{r}$$

Relationships between angular frequency, frequency and period of rotation

The description of circular motion is sometimes given in revolutions per second. This is the same as the frequency f of the motion.

The angle swept out by the radial line OP in 1s is ω, which is

$2\pi \times$ the number of revolutions per second

$$\omega = 2\pi f$$

ω is sometimes referred to as the **angular frequency**.

The **period** T is the time for one revolution, so

$$f = 1/T \qquad T = 2\pi/\omega$$

The relationships are particularly important in work on simple harmonic motion and when dealing with alternating currents.

Derivation of the equation for centripetal acceleration

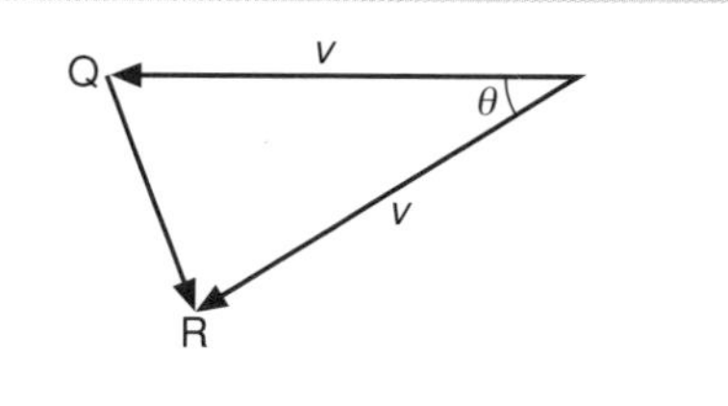

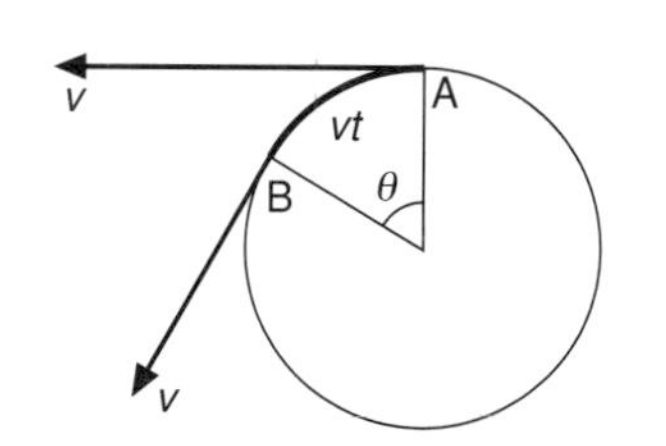

- In time t the particle moves from A to B at constant speed v.
- The particle moves along an arc of length vt in time t.
- The radial line sweeps out an angle θ in time t.

$\theta = vt/r$ (This will be used later)

The vector diagram for the change is shown above. The arrow QR represents the change in velocity:

Velocity change = $2v \sin(\theta/2)$

If we consider a short enough time that the velocity change is the instantaneous change in velocity, the angle will be a small angle so $\sin(\theta/2) = \theta/2$ (in rad).

Velocity change = $2v\theta/2 = v\theta$

Using the fact that $\theta = vt/r$,

$$\text{Velocity change} = \frac{vvt}{r} = \frac{v^2t}{r}$$

$$\text{Acceleration} = \frac{\text{velocity change}}{\text{time}} = \frac{v^2}{r}$$

Questions to try

1. A bicycle wheel of radius 0.25 m rotates 2.5 times per second. Calculate:
   (a) the angular speed in rad s^{-1}
   (b) the linear speed of a point on the outside edge of the tyre
2. Calculate the centripetal acceleration of a particle that is moving at a speed of 1.5×10^5 m s^{-1} in an arc of radius 0.18 m.
3. The angular frequency of a rotating object is 300 s^{-1}. Calculate:
   (a) the frequency of rotation
   (b) the period of rotation
4. Calculate, in rad s^{-1}, the angular speed of a yo-yo that completes 25 revolutions in 1.5 s.

If you want to know more about:
Radians see page 60
Vectors see pages 60–61

NON-LINEAR MOTION CONT: UNIFORM ANGULAR ACCELERATION

This is the equivalent of linear acceleration and is produced when there is an accelerating torque that increases the angular velocity. A torque may also decrease the angular velocity producing an **angular deceleration**.

Angular acceleration

Angular acceleration is the change in the angular speed divided by the time. It is given the symbol α.

If the angular speed changes in time t the angular acceleration is

$$\alpha = (\omega_2 - \omega_1)/t$$

The unit of angular acceleration is **rad s^{-2}**.

Equations for uniform angular acceleration

The equations relating angular terms are similar to those in linear motion:

$$\theta = \tfrac{1}{2}(\omega_2 + \omega_1)t$$
$$\omega_2 = \omega_1 + \alpha t$$
$$\theta = \omega_1 t + \tfrac{1}{2}\alpha t^2$$
$$\omega_2^2 = \omega_1^2 + 2\alpha\theta$$

- θ is the angular displacement in time t.
- ω_1 is the initial angular velocity.
- ω_2 is the final angular velocity.
- α is the angular acceleration.

Note: angular acceleration is not the same as centripetal acceleration. Angular acceleration occurs when the speed of rotation is increasing. There is centripetal acceleration even when the angular speed is constant.

Example

The radius of a video tape spool is 0.043 m. It accelerates to a rotational speed of 5.0 revolutions per second in a time of 4.0 s. Calculate:

(a) the angular acceleration
(b) the angle turned through during the accelerating time
(c) the centripetal acceleration at the edge of the spool at its maximum speed.

(a) angular speed = $2\pi 5$ = 31 rad s^{-1}
angular acceleration = change in speed/time
= 31/4.0 = 7.8 rad s^{-2}

(b) $\theta = \omega_1 t + \frac{1}{2}\alpha t^2$
$\theta = 0 + \frac{1}{2} \times 7.8 \times 4.02 = 62$ rad

(c) centripetal acceleration $= r\omega^2$
$= 0.043 \times 31^2$
$= 41$ m s^{-2}

Question to try

A proton in an accelerator accelerates from 2.0×10^5 m s^{-1} to 4.0×10^6 m s^{-1} in 1.75 s. The proton moves in a circular path of constant radius 250 m.

(a) Assuming that the acceleration is uniform calculate the angular acceleration of the proton.
(b) How far does the proton move while accelerating?
(c) Calculate the initial and final centripetal forces required.

PARABOLIC MOTION

When an object moves at constant velocity in one direction and has a constant acceleration at right angles to the direction of constant velocity, the path followed is a parabola. This is the case for the motion of a mass in a uniform gravitational field and a charged particle in a uniform electric field.

Independence of motion in two directions at right angles

When an object is moving in two dimensions, the motion can be analysed in two directions at right angles independently. The situations here have an acceleration in one direction only, but situations involving both directions can also be analysed in this way.

Which directions are used?

In the motion of an object in the gravitational field near the Earth's surface, the horizontal and vertical directions are the important directions.

For a charged particle, one direction is the direction of the initial velocity and the other is the direction of the uniform field.

Example 1: Motion of a projectile in a gravitational field

A heavy ball is thrown horizontally outwards at a speed of 18 m s^{-1} from the edge of the roof of a building that is 35 m high. How far will it travel horizontally before it hits the ground?

First use the vertical motion to find the time of flight. This is uniform acceleration at 9.8 m s^{-2}.

$s = 0$	use $s = ut + \frac{1}{2}at^2$
$u = 0$	$35 = 0 + \frac{1}{2}9.8t^2$
$v = ?$	$35 = 4.9t^2$
$a = 9.8$ m s^{-2}	$t^2 = 35/4.9 = 7.1$
$t = 0$	$t = 2.67$ s

Now consider the horizontal motion to find the distance required. This is uniform speed (assuming no air resistance effects). The time of flight is the same horizontally as vertically:

$s = vt$
$s = 18 \times 2.67 = 48$ m

Example 2: Motion of an electron in an electric field

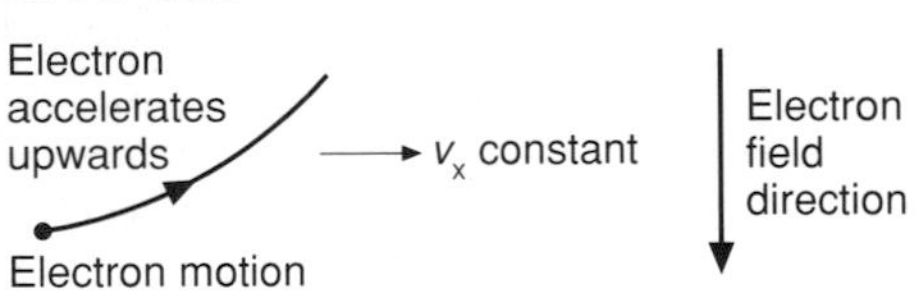

A high-speed electron travelling horizontally enters a region where it experiences a constant vertical acceleration a. Show that the path is a parabola.

The motion in each direction can be treated independently. Only time increases in the same way for both directions.

Horizontal speed is constant = v_x

In time t the horizontal distance travelled = $x = v_x t$

so $t = \frac{x}{v_x}$ *(This will be used later)*

The electron starts from rest vertically and accelerates uniformly. In the same time the electron will move a vertical distance y given by $ut + \frac{1}{2}at^2$

$u = 0$, so $y = \frac{1}{2}at^2$

To find the relation between x and y, t has to be eliminated. Using $t = \frac{x}{v_x}$, $y = \frac{1}{2}\frac{ax^2}{v_x^2}$

Since $\frac{1}{2}$, a and v_x are all constant,
$y = \text{constant} \times x^2$.
So the path is a parabola.

If you want to know more about:
Equations of linear motion see page 20
Radians see page 62

Logarithms

When are logarithms to base 10 (log) used?

1. Logarithms to the base 10 produce more manageable numbers when comparing very large and very small numbers. E.g. in determining pH values in chemistry and amplification in decibels in electronics.
2. Converting large numbers to their log values makes it easier to represent the quantities on a graph.
3. Logarithms to base 10 are used to analyse data. They are used in attempts to confirm a power relation. They can also be used to determine the power if it is unknown, so that an equation relating the two quantities can be found.

Note: Checking ln values

The ln of 100 is

$$\ln 100 = 4.605$$

This means that $2.72^{4.605} = 100$. Check this with your calculator using the method above.

Note: Logarithms and units

You can only find the logarithm of a pure number, so **a logarithm has no unit**.

To determine the log or ln of a quantity, the quantity first has to be divided by its unit to convert it to a pure number.

For example, if you wish to determine the log of a length l measured in metres, you will determine the log of (l/m). You should therefore write log (l/m) at the head of a column in a table of data or on the axis of a graph.

Note: Some useful relationships when using logarithms

- $\ln e = 1$
- $\ln 10 = 1$
- $\ln xy = \ln x + \ln y$
- $\ln (x/y) = \ln x - \ln y$
- $\ln x^n = n \ln x$
- $\ln e^n = n \ln e = n$

Similar relationships hold for logs.

Base 10 and base e

These are the two important bases for scientific work: to base 10 are just called **logarithms**; to the base e are called **natural logarithms**.

The value of e is 2.72 to 3 significant figures (e is a 'never ending' number like π).

What is a logarithm?

Logarithms are linked to indices. Using index notation,

$$100 = 10^2$$

We say that 2 is the logarithm of 100 to the base 10.

The logarithm of a number tells us to what power the base has to be raised to give the number itself. So

- 10 has to be raised to the power 2 to get 100.
- 10 has to be raised to the power 3 to get 1000, etc.

Calculating the log and ln of a number

- When you see **log** it means the base is 10.
- When you see **ln** it means the base is e.

These are the labels on the keys on your calculator. To determine the log or ln simply insert the number and press log or ln.

Check that you can use your calculator by confirming the log and ln of the following numbers.

	log	ln
3.54	0.549	1.264
0.000 267	−3.573	−8.228
2.45×10^{14}	14.389	33.132
1.73×10^{-8}	−7.762	−17.873

When you use logarithms do not include the number before the decimal point in your significant figure assessment. This will cause you to lose accuracy in your calculations and graphs.

When are logarithms to the base e (ln) used?

Many natural phenomena are related to another quantity by an exponential relationship. These are relationships that include base e in the laws.

1. ln values are best used in testing for exponential changes. They avoid the need to include the factor of 2.30.
2. Logarithms to the base e can be used to analyse data for power laws instead of log values.

What is the log of a number like 1300?

It is not easy to see the power of 10 that will give 1300. It happens to be 3.114. Using your calculator you can prove to yourself that

$$10^{3.114} = 1300$$

Using one type of scientific calculator you insert 10, press INV then x^y, insert 3.114 and press =.

Note: Relative sizes of log and ln of a number

The magnitude of the ln is always larger than the log by 2.30. Use the figures in the table to check this.

When x is a number,

$$\ln x = 2.30 \log x$$

What number corresponds to a given log or ln?

This is called finding the **antilog** of a number.

For base 10

- This is called the antilog (INV log on the calculator).
- If y is the log the antilog is simply the value of 10^y.
- From the table the antilog of 0.549 is 3.54.

For base e

- If y is the ln the number is e^y.
- From the table the INV ln of 33.132 is 2.45×10^{14}.

If you want to know more about:

Indices see pages 8–9
Analysing data see pages 67–68

Using logarithms

On this page we will look at how log values are used to provide more manageable numbers when comparing very large or very small numbers and to make it possible to represent a wide range of very large and very small numbers on a graph.

PLOTTING GRAPHS

Avoiding graph plotting problems

Physics: When testing resonant LCR circuits (containing both inductors and capacitors) in physics and amplifier gain in electronics the response has to be tested over a wide range of frequencies, e.g. from 10 Hz to 1 MHz. To plot the low frequencies on a linear scale on a typical sheet of A4 graph paper is impossible. If 200 mm were used to represent 1 MHz (1000 000 Hz), then 10 Hz would be represented by only 0.002 mm, 100 Hz by 0.02 mm and 10 000 Hz by 2 mm, so that results between 10 Hz and 10 000 Hz would be crammed into only 2 mm of the scale.

Biology: A similar problem arises in a biological test of how the number of bacteria in a culture varies with time. Numbers can range from relatively few at the start of an experiment to many thousands at the end.

Example: Output voltage V to be plotted against frequency f

f/ Hz	log (f/ Hz)	V
10	1.00	0.10
100	2.00	0.35
1 000	3.00	0.45
10 000	4.00	0.45
100 000	5.00	0.40
1 000 000	6.00	0.25

Graph of V against log (f/ Hz)

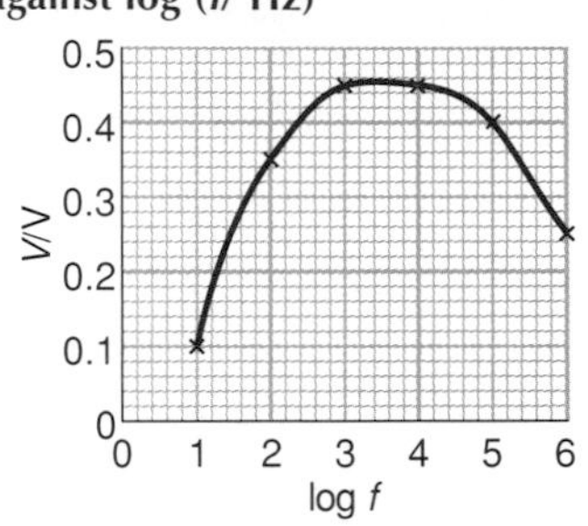

Graph of V against f

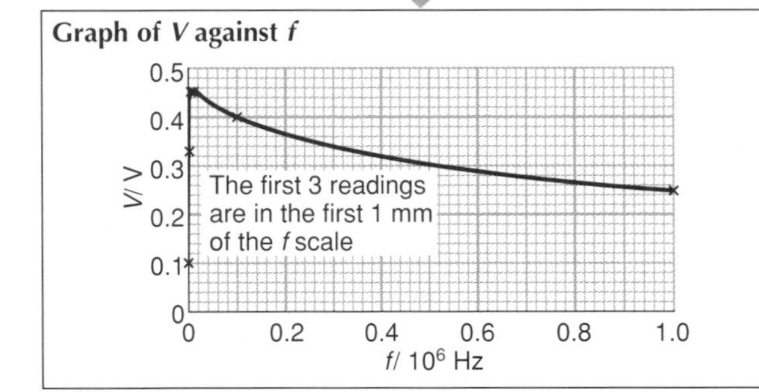

Note: Data are easily plotted on the log scale, showing the trend clearly. This is impossible on the linear scale. Intermediate values can be determined more easily using the log scale.

Note: Remember that you would need to find the antilog to find an actual frequency when reading from the graph.

MAKING NUMBERS EASIER TO MANAGE

pH values

The acidity or alkalinity of a solution is related to the concentration of hydrogen ions in it. The concentration of hydrogen ions ranges from 1 mol dm^{-3} (1 mole per litre) in a strong acid to 1×10^{-14} mol dm^{-3} in a strong alkali. Pure water is neutral and contains 1×10^{-7} mol dm^{-3} of hydrogen ions.

The wide range makes the numbers unmanageable so to make it simpler pH is used. The pH is defined as minus the log of the hydrogen ion concentration (written as $[H^+]$):

$$pH = -\log [H^+]$$

The – sign is included to make the pH positive.

- A strong acid has a pH of $-\log 1 = 0$.
- A neutral solution has a pH of $-\log (1 \times 10^{-7}) = 7$.
- A strong alkaline solution has a pH of $-\log (1 \times 10^{-14}) = 14$.

Note that a change of 1 in the pH value means a tenfold increase in hydrogen ion concentration.

Calculating pH

To calculate pH you need to know the hydrogen ion concentration.

- The pH of an acid that contains 0.2 mol dm^{-3} in an aqueous solution is $-\log (0.2) = 0.7$.
- For dibasic acids there are two hydrogen ions for each mole of acid in solution, so you have to double the concentration before taking the log.
- To find the pH of a base that contains 0.02 mol dm^{-3} in an aqueous solution proceed as follows:

Dissociation constant for water $K_W = [H^+][OH^-]$
$= (1 \times 10^{-7}) \times (1 \times 10^{-7})$
$= 1 \times 10^{-14}$ mol^2 dm^{-6}

This is constant. In a base that contains 0.02 mol dm^{-3},

$$[H^+] = \frac{1 \times 10^{-14}}{0.02} = 5 \times 10^{-13}$$

$$pH = \log (5 \times 10^{-12}) = 12.3$$

Note: The same principle is used to deal with the dissociation of acids and bases in water.

- For acids, $pK_a = -\log (K_a)$.
- For bases, $pK_b = -\log (K_b)$.

K_a and K_b are the dissociation constants.

Questions to try

1. Determine the log and ln values of the following numbers:
   (a) 0.234 (b) 2.34 (c) 2.34×10^{10} (d) 9.45×10^{-6}
2. Calculate the pH of:
   (a) 0.04 mol dm^{-3} HCl (aq) (b) 0.005 mol dm^{-3} HNO_3
   (c) 0.05 mol dm^{-3} H_2SO_4 (d) 0.04 mol dm^{-3} NaOH
   (e) 0.1 mol dm^{-3} KOH

Testing for an exponential relationship using log graphs

Practical work often has the objective of proving that a relationship between two quantities agrees with a given hypothesis (or theory) or finding an equation that relates two quantities. A log graph is a powerful tool in this task. If the variables in an experiment produce a curve when plotted directly on a graph, the variables **may** be related by a power law or by an exponential relationship. The tests on these two pages enable you to determine whether or not this is the case.

Exponential graphs?

If a graph or part of it looks like A the change may be an exponential decay; if it looks like B it may be an exponential growth. Note that it is possible for data to vary exponentially over only part of the range.

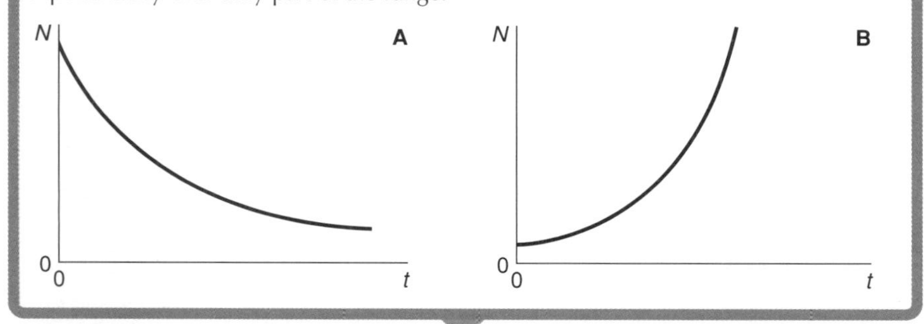

Testing if two quantities vary exponentially (q with p)

1. Determine the log or ln of the quantities for q.
2. Do nothing to the values for p.
3. Plot a graph of log q or ln q against p. That is, plot log q or ln q on the y-axis and p on the x-axis.

If the best line through the points is a straight line, q varies exponentially with p.

- A negative gradient shows that the change is an exponential decay.
- A positive gradient shows that it is an exponential growth.

Mathematical reasoning

The equation for an **exponential decay** with time would be

$$p = p_0 e^{-kt}$$

- p is the quantity that is changing with time t.
- p_0 is the value of the quantity at time $t = 0$ (since $e_0 = 1$).
- k is a constant for the decay.

Taking ln (logarithms to the base e) on both sides gives

$$\ln p = \ln p_0 + \ln(e^{-kt})$$
$$\ln p = \ln p_0 - kt$$

so, $\ln p = -kt + \ln p_0$.

Example: Growth of bacteria

In an experiment the number of bacteria in a culture was counted. The first count was 3 days after the culture was formed and other counts were made at irregular intervals:

Time/ days	3	6	10	12	13
Number of bacteria, N	170	303	630	960	1120

(a) Show that the number of bacteria increases exponentially with time.

(b) Determine the number of bacteria present when the culture was made.

The values of log N are determined and a graph of log N against time is plotted.

Time/ days	3	6	10	12	13
log N	2.23	2.48	2.80	2.98	3.04

3.2
3.0
2.8
2.6
2.4
2.2
2.0
1.8
log N
0 2 4 6 8 10 12 14
Time/ days

Interpreting the graph

The graph of log N against time is a straight line of positive gradient, showing that the bacterial growth was exponential over this period.

When $t = 0$ the intercept on the log N axis is 1.98. The INV log is 95 (to the nearest whole number) so there were 95 bacteria present at the time the culture was made.

Determining the equation for a change with time

The easier graph to use is the ln graph. Compare the ln equation with the general equation for a straight line:

$$\ln p = -kt + \ln p_0$$
$$y = mx + c$$

ln p is on the y-axis and t is on the x-axis.

- The **gradient** of the graph is $-k$. The unit for k is s^{-1}.
- The **intercept** on the ln p (y-axis) is ln p_0. Calculating the INV ln of the intercept gives p_0.

Note: Can log be used instead of ln?

Yes it can. The graph will still be linear for an exponential change but the gradient and intercept have different meanings. For a graph of log q against p,

$$\text{Gradient} = \frac{\text{decay constant}}{2.30} = \frac{-k}{2.30}$$

$$\text{Intercept} = \log p_0$$

Testing for a power law

What is a power law?

When two quantities p and q are related by a power law the equation relating them is $p = Aq^n$

In this equation A and n are constants.

- The constant A may be a positive or negative number but in practical situations it is usually positive.
- The constant n may also be positive or negative and may be a fraction or a decimal.

Although in many situations the value of n is a number such as ± 1/2, ± 1, ± 2, the number may be any value.

Example

The following data were obtained in an experiment to investigate how the count rate (corrected for background radiation) C varied with distance d from a radioactive source.

d/ cm	3.0	4.0	5.0	6.0	7.0
C/ s^{-1}	56	32	20	14	10

The count rate is expected to obey a power law.

(a) Show that the data are consistent with a power law.

(b) Determine the equation that relates count rate C to distance d from the source.

- First find the logs of C and d.

log (d/ cm)	0.477	0.602	0.700	0.778	0.845
log (C/ s^{-1})	1.78	1.51	1.30	1.15	1.00

- Now plot the graph.

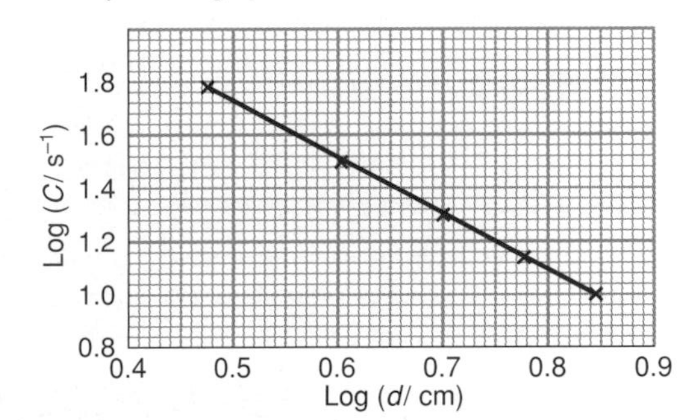

To test whether q is related to p by a power law

1. Determine the log or ln of **both** p and q. Do not take the ln of one and the log of the other. You must do the same for both.
2. Plot a graph of log q against log p or ln q against ln p. That is, plot log q or ln q on the y-axis and log p or ln p on the x-axis.

If the best line through the points is a straight line, the relationship between the two quantities is a power law.

- A negative gradient shows an inverse proportionality: $q \propto 1/p^n$.
- A positive gradient shows direct proportionality: $q \propto p^n$.

Using some of the useful relationships on page 65 and taking logarithms of both sides of the equation, $p = Aq^n$, gives

$$\log p = \log (Aq^n)$$
$$\log p = \log A + \log (q^n)$$
$$\log p = \log A + n \log q$$

or, rearranging,

$$\log p = n \log q + \log A$$

Note that since A is constant its log is constant.

Note: Identifying units in log graphs

It is only possible to find the log of a pure number, and when a log is plotted it is itself a pure number. This is a subtle point but it does mean that you have to be careful when labelling the axes.

For example, in one experiment a distance d is measured in metres. If the log is to be plotted, the correct axis labelling is log (d/m).

If the value were 5.0 m, the plotted point would be log (5.0 m/m) which is the log of 5.0 (i.e. 0.699).

Interpreting the graph

The best line through the points is a straight line with a negative gradient so that the data are related by an inverse power law. The equation should have the form $C = Ad^n$ where n will be negative.

- The gradient is $-\frac{0.62}{0.30} = -2.1$, so the law is $C \propto 1/d^{2.1}$.
- The intercept on the log (C/s^{-1}) axis is 2.78.
- The value for the constant A is 603 s^{-1}.

The equation relating C and d from this data is therefore

$$C = 603d^{-2.1} = \frac{603}{d^{2.1}} \quad \text{or rounding off} \quad C = \frac{600}{d^2}$$

(The constant 2.1 is close to 2 so an inverse square law is a distinct possibility.)

Determining an equation for a power law

Plot the graph using either ln or log values. It is assumed here that the graph is the log graph.

Compare the log form of the equation with the general equation for a straight line.

$$\log p = n \log q + \log A$$
$$y = mx + c$$

- log p is plotted on the y-axis.
- log q is plotted on the x-axis.

Gradient of graph $= n = \frac{\text{change in } \log p}{\text{change in } \log q} = \frac{\Delta(\log p)}{\Delta(\log q)}$.

The logs do not have units so the gradient has no units.

The **intercept on the log p axis** is **log A**. Calculating the INV log of the intercept gives A.

Take care to notice whether or not there is a false origin when determining the intercept. This is common in this type of graph.

Questions to try

1. The table shows the intensity I of gamma radiation that passes through absorbers of different thickness as a percentage of the original intensity. Theory suggests that the equation for the intensity is $I = I_0 e^{-\mu t}$ where t is the thickness of the absorber and μ is a constant.

   (a) Show that the data agreed with this theory.

   (b) Determine the value for μ for the absorber used.

   (c) Determine the thickness that reduces the intensity to half the original value.

t/ m	0	0.020	0.040	0.060	0.080	0.100	0.120
I/ %	100	73	54	40	29	21	16

2. The following data relate the radius R of a nucleus to the nucleon number A of the nucleus. The relation between them is thought to obey the law $R = R_0 A^n$. Analyse the data to check that this is the case and determine the values of R_0 and n.

Element	C	Fe	Br	Ba	W
A	12	56	81	138	184
R/ 10^{-15}m	3.2	5.3	6.0	7.2	8.0

Probability

Probability is a way of expressing the chance of an event occurring. There are numerous events in science that occur because of chance. Understanding probability is also important in assessing the data obtained in experimental science.

Probability in throwing dice

When you throw a die the chance of getting a 6 is 1 in 6: the probability is 1/6 or 0.167. The probability of not getting a 6 is 5/6 = 0.833. There are five ways of not getting a 6 and only one way of getting it. This gives rise to an important rule.

What is most likely to happen is what can occur in the greatest number of ways.

In the case of the die, it is most likely that you will not throw a 6.

This happens in nature, too. An object placed in contact with a colder object is always seen to give up energy. This is because there are more ways of sharing the energy when this happens. If the colder object got colder and the hot body hotter the energy would be shared out in a smaller number of ways, so this is less likely to happen by chance.

Throwing two dice

When you throw two dice the chance of getting two 6s is 1 in 36. This is also true for all the other doubles. The table shows how this occurs. In the shaded cell die 1 shows a 5 and die 2 shows a 4. The totals are shown in brackets.

The table shows the 36 possible combinations. 1–1 and 6–6 appear only once. They each have a probability of occurrence of 1/36 = 0.0278.

The most likely total is 7. This occurs 6 times, and so has a probability of occurrence of 0.167.

Die 2 \ Die 1	1	2	3	4	5	6
1	1–1 (2)	1–2 (3)	1–3 (4)	1–4 (5)	1–5 (6)	1–6 (7)
2	2–1 (3)	2–2 (4)	2–3 (5)	2–4 (6)	2–5 (7)	2–6 (8)
3	3–1 (4)	3–2 (5)	3–3 (6)	3–4 (7)	3–5 (8)	3–6 (9)
4	4–1 (5)	4–2 (6)	4–3 (7)	4–4 (8)	4–5 (9)	4–6 (10)
5	5–1 (6)	5–2 (7)	5–3 (8)	5–4 (9)	5–5 (10)	5–6 (11)
6	6–1 (7)	6–2 (8)	6–3 (9)	6–4 (10)	6–5 (11)	6–6 (12)

Examples of probability in science

Radioactive decay: The half-life of radioactive materials is related to the probability of a nucleus decaying each second.

Wave–particle duality: The amplitude of the wave associated with a particle tells us the about the probability of finding a particle at that point.

Reaction rates: Reaction rates in chemistry are related to the chance of two atoms with the right energy meeting. The higher the probability the higher the reaction rate.

Assessing data in biology investigations: The outcomes of biological investigations are tested to see whether what has been observed is any more than a chance occurrence.

Our own characteristics are the result of chance: Experiments in genetics have shown that the frequency of particular characteristics in offspring is governed by chance.

Probability in tossing a coin

When you toss a coin it may show a head or a tail.

- The chance of seeing a head is 1 in 2. We say that the probability is $\frac{1}{2}$ and usually express this as a decimal: the probability of the coin showing a head is 0.5.
- There is clearly also a 0.5 chance of it showing a tail.

The total probability of all events must add up to 1 because one of the events must happen.

Tossing two coins

When you toss two coins at the same time, each coin has a 0.5 probability of being a head (H) and 0.5 probability of being a tail (T).

- Probability of HH = 0.5 × 0.5 = 0.25.
- Probability of TT = 0.5 × 0.5 = 0.25.

HT can occur in two ways, HT or TH.

- Probability of HT = 2 × 0.25 = 0.5.

Probability in genetics

The same principles for assessing the likely outcome can be applied in predicting the outcome in genetics experiments. Instead of heads and tails the options are governed by the alleles (types of gene).

A is a dominant gene and **a** a recessive gene (see page 69). The probability of an outcome with a dominant gene present (AA, Aa or aA) is 0.75 and the probability of two recessive genes is 0.25.

	A	a
A	AA	Aa
a	aA	aa

In Mendel's experiment **A** was the allele for a tall plant and **a** the allele for a short plant. Tall was dominant, so the outcome was that 75% of the offspring were tall plants.

Combining two characteristics

When two characteristics combine, the situation is more complex. In experiments using peas the characteristics of the peas could be yellow or green and round or wrinkled.

The round (R) and yellow (Y) alleles were dominant, so when these appear in the offspring that's what you get.

Round green (Ry) combining with wrinkled yellow (rY) produces round yellow offspring. The possible combinations are shown below.

Male \ Female	RY	Ry	rY	ry
RY	RR–YY	RR–Yy	Rr–YY	Rr–Yy
Ry	RR–Yy	RR–yy	Rr–Yy	Rr–yy
rY	Rr–YY	Rr–Yy	rr–YY	rr–Yy
ry	Rr–Yy	Rr–yy	rr–Yy	rr–yy

The offspring have characteristics in the ratio 9 : 3 : 3 : 1. The probability of a plant having round yellow peas was 9/16 or 0.56 and the probability of it having both recessive alleles, leading to wrinkled green peas, was 1/16 or 0.063.

If you want to know more about:
Evaluating experimental data see pages 76–89
Probability in genetics see page 70

The Hardy–Weinberg equation

What is the Hardy–Weinberg equation used for?

It is used to calculate the frequencies of alleles and genotypes in a population, i.e. to calculate the percentage of organisms (animals or plants) that have a particular characteristic and are therefore carrying a particular gene.

The Hardy–Weinberg equation is:

$$p^2 + 2pq + q^2 = 1$$

where $p + q = 1$.

In a given population

- p is the frequency of occurrence of allele **A**
- q is the frequency of occurrence of allele **a**.

The frequencies are expressed as decimals. For example, if a particular allele **A** is present in 30% of the population, the value is 0.30.

$(p + q)$ must equal 1 because the allele must be **A** or **a**.

Where does the Hardy–Weinberg equation come from?

When mating takes place there is random fertilization.

- p will be the frequency of the allele **A** in the female population.
- p will also be the frequency of the allele **A** in the male population.

The result is that p^2 will be the frequency of the **AA** genotype in the offspring, i.e. the probability of a female gamete with genotype **A** meeting with a male gamete of genotype **A** is p^2.

The frequency of each genotype is:

		Female gametes	
		A (p)	a (q)
Male gametes	A (p)	AA (p^2)	Aa (pq)
	a (q)	Aa (pq)	aa (q^2)

- Probability of **AA** $= p^2$ = frequency of **AA**.
- Probability of **Aa** $= 2pq$ = frequency of **Aa**.
- Probability of **aa** $= q^2$ = frequency of **aa**.

These terms must add up to 1, so

$$p^2 + 2q + q^2 = 1$$

Note: Assumption

The Hardy–Weinberg equation assumes that the frequency of occurence of an allele remains unchanged from one generation to another. (This is called genetic equilibrium.)

Note: Definitions

- A **genotype** is the genetic make-up of an individual organism with respect to the alleles under consideration.
- A **gene** is the basic unit of inheritance. A particular gene is responsible for a particular characteristic, e.g. the colour of a flower.
- An **allele** is one of a number of forms of the same gene which determines different characteristics. A flower allele may be associated with red or white, for example.
- A capital letter **A** represents a **dominant allele**. A dominant allele influences the characteristics of an individual even when there is an alternative allele present.
- A small letter **a** represents a **recessive allele**. A recessive allele will have an influence on the new generation only if there is another identical allele present.

Note: Heterozygous and homozygous organisms

Heterozygous organisms carry a genotype consisting of a dominant and a recessive allele (**Aa**).

Homozygous dominant organisms carry a genotype consisting of two dominant alleles (**AA**).

Homozygous recessive organisms carry a genotype consisting of two recessive alleles (**aa**).

Example

What is the outcome of the random mating of a population of males in which 50% possess genotype **Aa**, with a population of females in which 50% have genotype **Aa**?

$$p = q = 0.5$$

Therefore

$$p^2 = 0.50 \times 0.50 = 0.25$$
$$2pq = 2 \times 0.50 \times 0.50 = 0.50$$
$$q^2 = 0.50 \times 0.50 = 0.25$$

In the offspring there will be

- 25% **AA** (homozygous dominant genotype)
- 50% **Aa** (heterozygous genotype)
- 25% **aa** (homozygous recessive genotype).

Question to try

The frequency of rhesus-negative blood (genotype **rr**) in a population is 0.18. Those with rhesus-positive blood have the genotype **Rr** or **RR**.

(a) Calculate the frequency of the **r** allele in the population.

(b) What proportion of the population have the genotype **Rr**?

If you want to know more about:

Probability see page 69 

Frequency distributions

In science, all data are gathered by counting or measuring some quantity. Drawing frequency distributions of data often makes patterns easier to see and is a good way of communicating data visually.

Discrete data

In a biology investigation you may count the number of people with different blood groups or the number of plants bred from the same parent seeds that have flowers of a different colour. This type of data is **discrete** data. There will be no disputing the actual value (unless you have miscounted).

Bar charts

It is often useful to present data in the form of a bar chart to obtain a visual representation of the data. The bar chart shows the **frequency distribution**. This is often simply called the **distribution**. This tells you at a glance how often a particular value was obtained or the number of measurements in a given range. This type of bar chart is called a **histogram**.

For discrete data you record the frequency of occurrence of each characteristic. You would arrive at a bar chart like this:

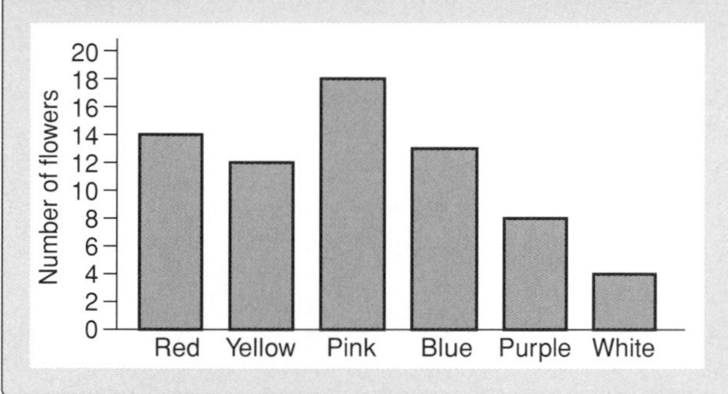

Example

In an investigation of heights of a group of students you may decide to use a class interval of 0.02 m (2 cm).
Having obtained your data (or even while collecting it) you make up a table of class intervals that includes the extreme values of your data.

For each value indicate with a stroke the class in which it lies. This gives a table like this:

Class interval/ m	Number of students
1.500–1.519	III
1.520–1.539	ⵁ I
1.540–1.559	ⵁ
1.560–1.579	ⵁ III
1.580–1.599	ⵁ ⵁ II
1.600–1.619	ⵁ II
1.620–1.639	IIII
1.640–1.659	ⵁ
1.660–1.679	II
1.680–1.699	I

Note that the class interval 1.520–1.539 m would include someone of height 1.539 m but not someone of height 1.540 m. (This assumes measurement to the nearest mm.)

Note: For both discrete and continuous data **the more measurements you have the more meaningful your data will be**.

Continuous data

You may measure the length of a leaf, a quantity of gas or the extension of a rod. You may make this measurement once or a number of times. Alternatively, you may measure the lengths of a number of different leaves, or the breaking strength of a number of different samples of the same thread.

In each case the measurements will fall within a range of possible values. This type of data is **continuous** data. Because of uncertainties in the data a single measurement could have any value in a given range. If 30 students made the same measurement each measurement would fall within a certain range of values and some measurements would probably be the same.

Drawing frequency distributions

For continuous data you first have to decide the **class interval**. This determines the ranges into which you are going to divide the data. This choice has to be made carefully. If you make it too large you may miss some interesting variations in the frequencies. Make it too small and you may not have enough data to see any differences.

You may need to try different groupings to find one that works, but a visual examination of the sample size and the range often gives you enough information to make a sensible judgement.

A histogram

This bar chart shows the frequency distribution of some data.

- The classes are shown at regular intervals on the *x*-axis.
- The heights of the columns represent the number of individual measurements in that class interval.

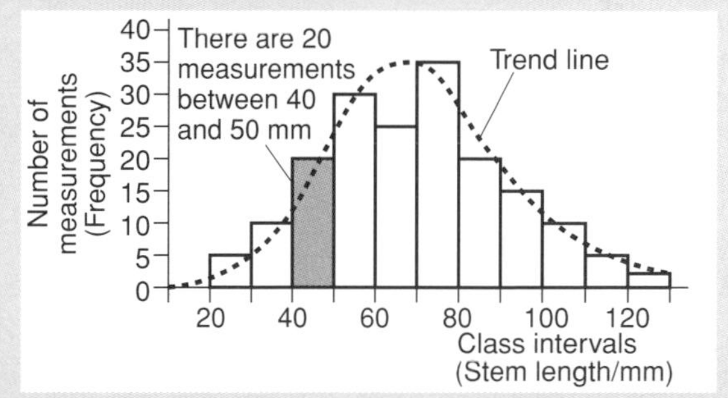

A curve may be added to show the general trend of the data.

The height of each bar represents the number in the population that fall in that class interval. Adding the heights of all the bars in the bar charts gives the total population in the sample.

Question to try

For the data in the example on the left:
(a) Draw the bar chart on the left.
(b) Determine the total population.
(c) In what class interval does the maximum height occur?
(d) What percentage of the total population has a height in this interval?

If you want to know more about:
Percentages see page 18
Using frequency distributions see pages 72, 74
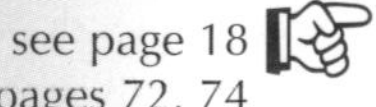

Shapes of distributions

It is interesting to inspect the way data are distributed. If you have drawn a bar chart the general shape of the distribution will be clear provided you have sufficient data. The distribution can then be drawn as a smooth curve. Part of any investigation in science is to discover reasons for the shape of an experimental distribution.

Unimodal distributions

Most distributions have only one maximum. These are called unimodal distibutions.

Normal distribution

The normal distribution is a unimodal distribution that occurs frequently in practice. It is symmetrical.

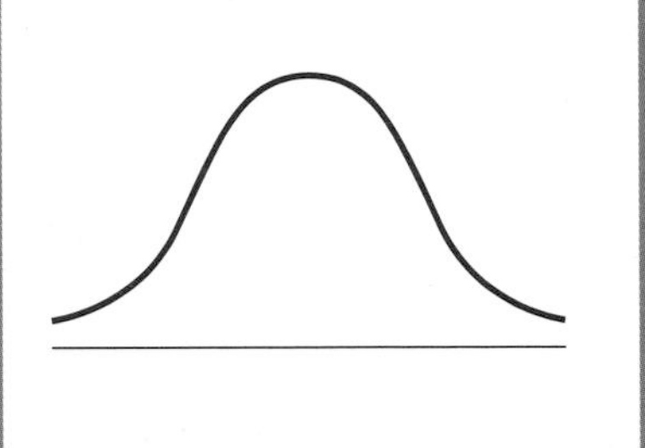

Skewed distributions

These are also unimodal distributions, but they are not symmetrical. The peak is not at the centre of the distribution curve. The curve shows a longer tail either to the right or to the left.

Skewed distributions may occur by chance when the sample is too small, even when a large sample would give a normal distribution.

They may also be representative of the true distribution. There may be some factor that has caused the skew.

This distribution is **positively skewed**. It is skewed to the left. The long tail is on the positive side of the distribution.

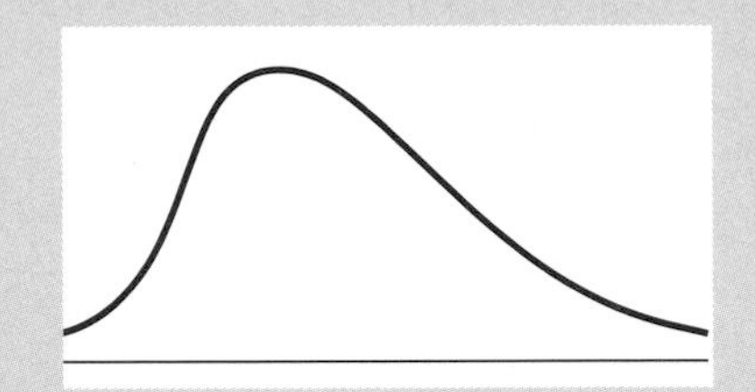

This distribution is **negatively skewed**. It is skewed to the right. The long tail is on the negative side of the distribution.

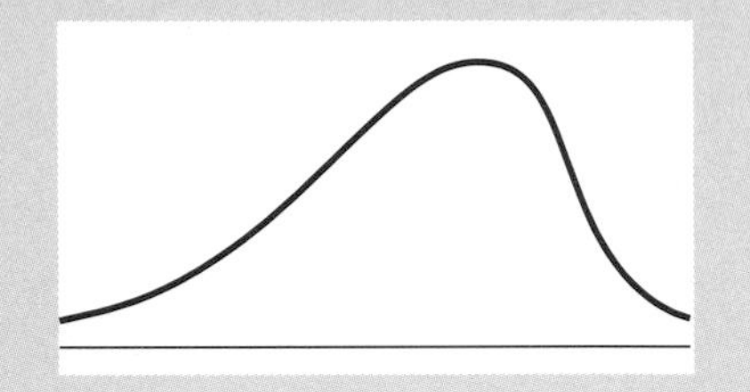

Bimodal distributions

A bimodal distribution has two distinct peaks. This may be because there are two distinct populations (groups) in the sample of data.

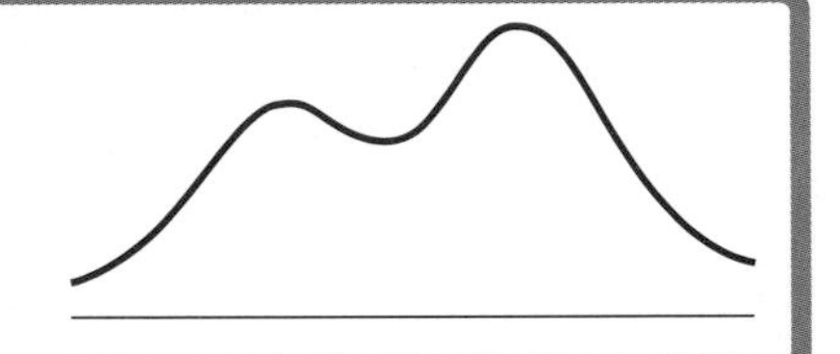

Note: If you find a skewed or bimodal distribution in your data, this needs to be explained in your analysis of the results.

Probability and distribution curves

When you draw a bar chart you have a **sample** of all possible events. If you could make all possible measurements you could state the precise frequency of any one of the events occurring. In practice you have only sample data, so you cannot say with certainty what the frequency is but you can give a good estimate – the probability. The larger your sample the better your estimate will be.

The purpose of statistical tests is to place a **confidence limit** on how good that estimate is.

Suppose you made 50 observations and produced this bar chart.

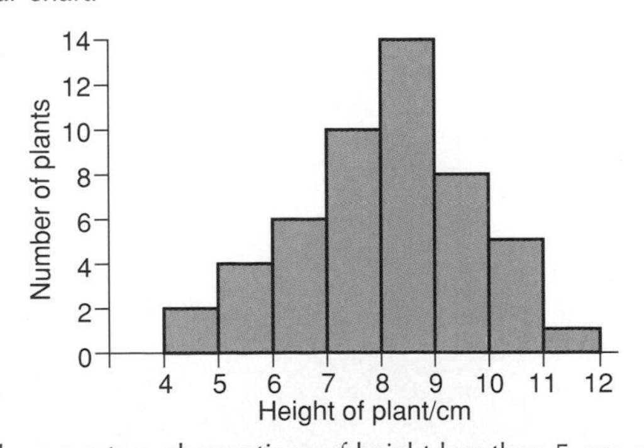

There are two observations of height less than 5 cm, so you can say that 2 in 50 or 4% of heights measured are lower than 5 cm. On the basis of your evidence you could suggest that there is a probability of 2/50 or 0.04 that a plant will be less than 5 cm tall.

Area under a distribution curve

The area under the distribution curve represents the whole population. The ratio

$$\frac{\text{area under curve within a given range}}{\text{total area under the curve}}$$

gives the proportion of population that falls within that range. If you select a specimen at random this represents the chance of that individual being within this range.

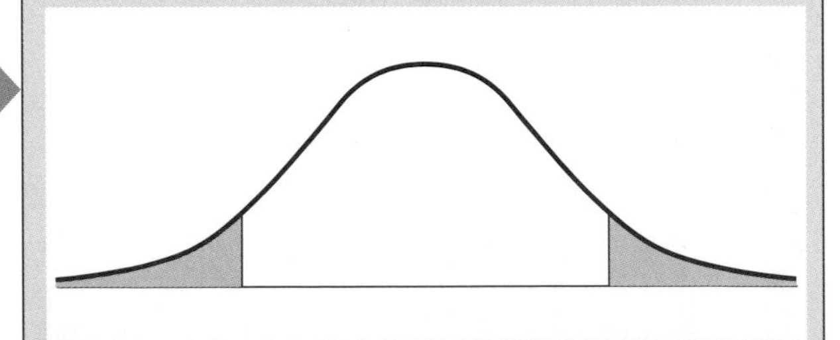

The shaded areas represent the proportion of the population that falls outside the range limits.

The ratio of these areas to the whole area is the chance of an observation being outside the limits.

Median, mean, and mode

The frequency distribution displays all individual measurements in the sample. This is the most complete display of the data. However, for many purposes, **statistical data** are used. The mean, median and mode are three types of statistical data. They are all ways of expressing an **average** for the data.

Median

This is the middle number in the list when all the measurements are arranged in rank order. This means that half the measurements are above the median and half below it.

If there is an even number of measurements the median is the arithmetic mean of the middle two numbers.

Note: If a sample is symmetrical about the centre, the mean and the median are identical. This is the case in a **normal distribution**.

Which is best?

Both the median and the mean are fair measures of the centre of a set of data.

The mean takes account of all the information provided by the data but can be influenced a lot by a few measurements that are a long way from the peak value (when there is a long tail in the distribution).

The median is useful in some clinical trials where the average survival time of treated patients cannot be determined until the last patient has died. The median survival time can be determined when half (+1) of the treated patients have died.

Mean

This is the arithmetic mean and is written $\bar{x}$ or <x>. The first is most common in statistical equations but the second form is sometimes used, in kinetic theory of gases for example.

$$\text{Mean} = \frac{\text{sum of all the measurements}}{\text{number of measurements}}$$

This is what we usually refer to when we talk about an average. In mathematical notation this is written

$$\bar{x} = \frac{\Sigma x}{n}$$

Differences

The difference between the mean and the median gives a measure of the degree of skew in the distribution.

The mean and the median move toward the tail of a skewed distribution. The mean usually moves the most.

This diagram shows the long tail and the mean, median, and modal values.

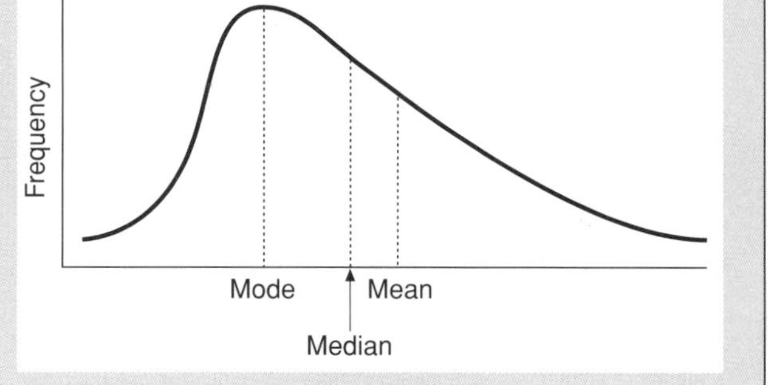

Mode

The mode is the measurement that occurs the greatest number of times. It will be the one that has the highest bar in the bar chart.

There can be more than one value for the mode; that is, two (or more) bars of equal height in the bar chart.

Example

The following is a list of measurements for the heights, in cm, of different specimens of the same flowers. The list is given in rank order.

7.1 7.3 7.4 7.4 7.5 7.5 7.5 7.6 7.6 7.6 7.6 7.6
7.8 7.9 7.9 7.9 8.1 8.1 8.2 8.3 8.4 8.4 8.5 8.6

Determine the mean, median and mode of the sample.

Number of measurements made = 24

$$\text{Mean height} = \frac{\text{sum of measurements}}{\text{number of measurements}} = \frac{187.8}{24} = 7.8 \text{ cm}$$

The list is even. The two central measurements are 7.6 (the 12th) and 7.8 (the 13th).

The median height is therefore the mean of these = 7.7 cm.

The mode is the most frequent value. This is 7.6 cm.

Question to try

The distribution shown is for the weights of plums taken from the same tree.

(a) How many plums were in the sample?
(b) Determine the mean, median and mode for the sample.

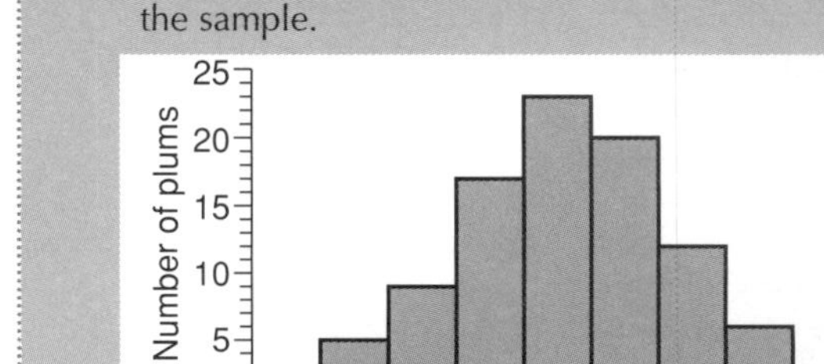

Note that the total weight of plums in a given class interval is the number of plums × the central (mean) value of the class interval.

If you want to know more about:
Frequency distributions see page 71 

The normal distribution and standard deviation

Many frequency distributions in scientific work look bell shaped. The bell shape is the characteristic shape of the **normal distribution**. Examples include different independent measurements (mass, length, etc.) of the same object and dimensions or masses of biological specimens of the same type, e.g. the leaves from a tree.

Normal distribution

In a normal distribution the curve (i.e. the general shape of the bar chart) is symmetrical about the central value.

A normal curve showing one and two standard deviations. The area under the graph is proportional to the number of measurements.

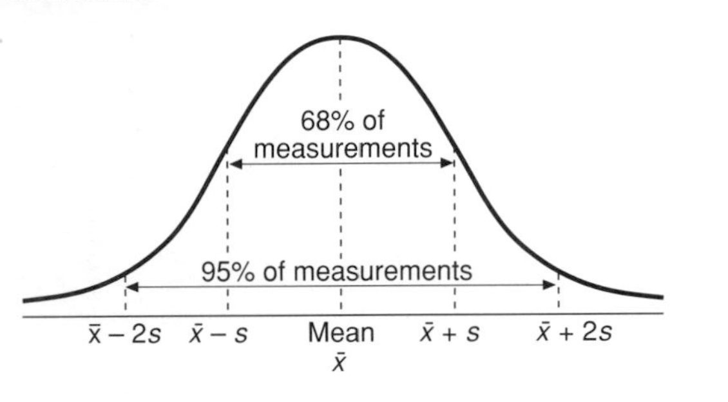

Note that, in practice, real data will only be approximately bell shaped. The more data you collect the nearer you will get to the normal distribution curve.

In a normal distribution 68% of the data are within 1 standard deviation *s* of the mean value of the sample.

68% of all the measurements lie between $\bar{x} + s$ and $\bar{x} - s$.

95% of the data are within **2 standard deviations** of the mean.

Probability and the normal distribution

Suppose that you are making measurements of a quantity that gives rise to a normal distibution. If you make a single reading there is a 68% **probability** that this single reading will fall within 1 standard deviation of the mean and a 95% chance that it will fall within 2 standard deviations.

Standard deviation

To calculate the standard deviation *s*:

1. Square each measurement.
2. Add the squared values together (Σx^2).
3. Divide by the number of measurements made ($\Sigma x^2/n$).
4. Square the mean value of the measurements ($\bar{x}^2$).
5. Subtract the result of step 4 ($\bar{x}^2$) from the result of step 3 ($\Sigma x^2/n$).
6. Take the square root. This is the standard deviation.

Written as a mathematical equation the standard deviation *s* is given by

$$s = \sqrt{\left(\frac{\sum x^2}{n} - \bar{x}^2\right)}$$

Note: Strictly, since the measurements are usually only a sample of all possible measurements, the standard deviation of the sample population is given by

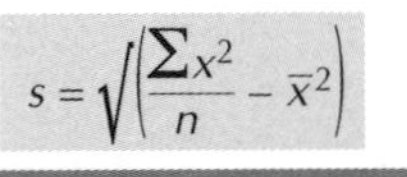

$$s = \sqrt{\left(\frac{\sum (x - \bar{x})^2}{n - 1}\right)}$$

Note: Using a calculator

The process may look daunting and time consuming. You may need to know how to carry out the process as part of your course, but in practical work you can take advantage of technology and make use of a scientific calculator. In 'statistics mode' the calculator does all the work for you so that you only need to interpret what the figures mean. It is worthwhile learning how to use a calculator for this purpose if you are going to do a lot of statistical work. This is covered in detail on the next page.

Example: Calculating the standard deviation

The table shows the masses of 10 babies born in a maternity ward. Calculate the mean mass of the babies and the standard deviation from the mean.

Measurement number	Mass/ kg	Mass^2/ kg^2
1	2.95	8.70
2	3.06	9.36
3	3.40	11.56
4	3.11	9.67
5	3.15	9.92
6	3.30	10.89
7	3.32	11.02
8	3.26	10.62
9	3.15	9.92
10	2.98	8.88
Σ	31.68	100.54

1. The mean mass = 31.68/10 = 3.168 kg
2. The 'mean mass' squared = 10.036
3. The 'sum of the masses^2' divided by 10 = 10.054 kg^2
4. The difference between these two values is

   10.054 – 10.036 = 0.018 kg^2

5. The square root of 0.018 = 0.134 kg

We can therefore say that the mean mass of the sample of 10 babies was 3.168 kg with a standard deviation of 0.134 kg. This means that there would be a 95% chance of a baby being born in the range 3.168 ± 0.268 kg (2.90 kg to 3.44 kg).

Questions to try

1. Calculate the mean and standard deviation of the following measurements for the diameter, in mm, of a glass rod that was measured using a micrometer.

   6.24 6.31 6.38 6.34 6.27 6.33 6.34 6.30

2. Determine the standard deviation for the measurements of flower heights in example 1 on page 73.
3. Calculate the standard deviation in the weight of plums in the question to try on page 73.

Using a calculator to determine statistical data

Some scientific calculators can operate in a statistics mode, which takes a lot of the drudgery out of statistical work and leaves you more time to interpret the data. The boxes here show how values are entered and the statistical data retrieved for one type of scientific calculator.

Entering data

1. The calculator is set into statistics mode. The handbook tells you how to do this. In my particular calculator this is done by first pressing the MODE key then 3.
2. The display indicates that this has been done correctly (e.g. SD is displayed).
3. Previous data are cleared by pressing the INV key (the alternate function key) followed by the key marked $\frac{\text{AC}}{\text{KAC}}$
4. Values are now entered. Enter the first item; press the DATA key; enter the second item; press the DATA key; and so on.

Retrieving statistical information

1. To obtain the number of entries, press K_{out} and then n (n is the number 3 key).
2. To obtain the mean, press INV followed by $\bar{x}$ ($\bar{x}$ is the 1 key).
3. To obtain the standard deviation for the sample, press INV followed by 3.
4. To obtain the standard deviation for the population, press INV followed by 2.

Other information can be retrieved using other keys.

Standard error and confidence limits

Standard error of the mean

Because a standard deviation is derived from a relatively small sample of all possible measurements, the mean and the standard deviation are estimates of the mean and standard deviation of the whole population.

There is *a sampling error*. A calculation of the standard error (SE) takes account of this. The complete term is **standard error of the mean**, SEM. It enables you to suggest the range in which the true mean of the total will fall.

It is calculated by dividing the standard deviation by the square root of the number of readings:

$$\text{Standard error} = \frac{s}{\sqrt{n}}$$

The standard error is important because it is used to define **confidence limits**.

Difference between standard deviation and standard error

- 95% of the data fall within 2 standard deviations of the mean.
- When using standard errors the mean falls in the range ± 1.96 SE.
- When using confidence limits this is called the $p = 0.05$ level.

Example

Refer to the data in example 1 on page 74.

(a) Determine the standard error.

(b) What are the 95% confidence limits and what information does this give?

$$\text{Standard error} = \frac{s}{\sqrt{n}} = \frac{0.134}{\sqrt{10}} = \frac{0.134}{3.16} = 0.042 \text{ kg}$$

The 95% confidence limits are (3.168 + 0.082) kg and (3.168 – 0.082) kg. That is 3.250 kg to 3.086 kg.

You could be 95% confident that the mean mass of all babies born in the ward would be in this range.

Confidence limits

How confident can you be that the mean you find from a sample of measurements represents the true mean?

This is expressed by a **confidence interval**. The limits of the confidence interval are the **confidence limits**.

95% confidence limit = mean value ± 1.96 SE

This means that you are 95% confident that the true mean value lies in this range. There is therefore only a 5% chance that the true mean value falls outside this range.

In your field work a 95% confidence limit will usually be adequate. In clinical trials (of the success of a drug for example) the confidence limits are wider because there is a need to be more certain of the outcome: 99.9% or even 99.99% may be used.

Critical values

In a statistical test of data you obtain a number that is derived from the data. This number provides information about the data that you have collected. The critical value tells you the value you need to have calculated to reach the appropriate confidence limit. This confidence limit is found from a table.

Other confidence limits

Tables enable other confidence limits to be found:

- For 99% confidence, the limits are ± 2.58 SE ($p = 0.01$ level).
- For 99.9% confidence, the limits are ± 3.30 SE ($p = 0.001$ level).

Note: What does $p = 0.05$ level mean?

The $p = 0.05$ level means that there is only a **5% probability** that the value lies outside the limit set by this level. This corresponds to a **confidence limit of 95%**. You can say that you are 95% confident that the mean value is between the two limits and there is only a 5% chance that it is outside these limits.

Comparison of means test

When might this test be used?

1. You will have collected two different sets of data that you want to compare.
2. Each set of data will have **at least 30** individual items in it and preferably a lot more.
3. You will probably have drawn bar charts for the sets of data, which suggest that there are two distinct sets of data. You need a statistical test to make a confident conclusion.

1: Indistinct sets of data
What do you suggest when the data for the two sets look like this?

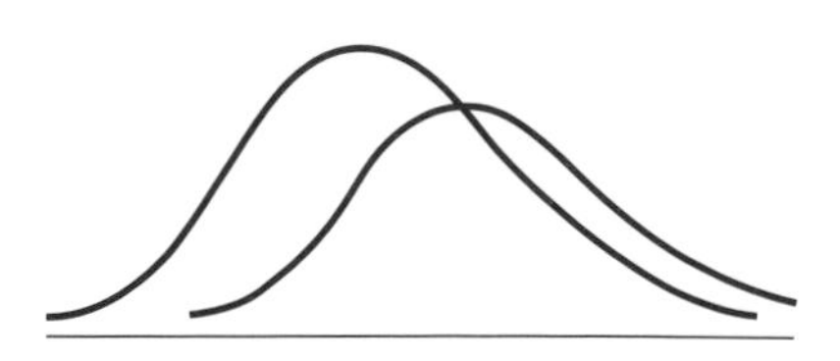

This could represent two different populations or you may just have happened to select a different sample of the same population by chance. You need to undertake a statistical test.

2: Distinct sets of data
If you plot two sets of data on a common class axis you may obtain distributions like these.

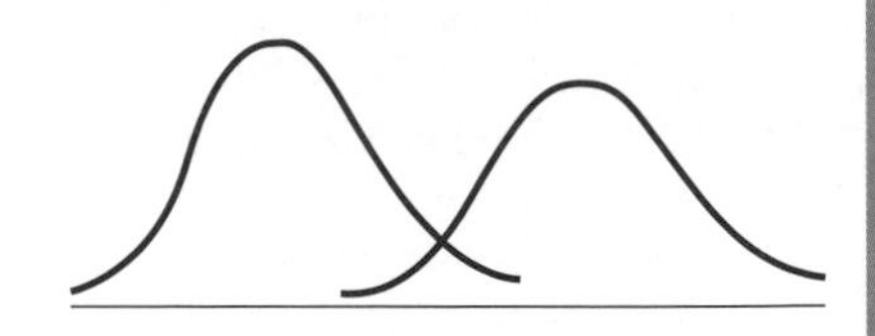

You would be justified in suggesting that these sets of data represented two different populations. No further test is necessary.

Carrying out the test

1. Calculate the mean of each sample.

$$\text{Mean} = \frac{\Sigma x}{n}$$

2. Calculate the difference between the two means.
3. Calculate the square of the standard deviation s^2 for each set of data:

$$\text{Standard deviation} = s^2 = \frac{\Sigma x^2}{n} - \bar{x}^2$$

4. Calculate the standard error in the difference between the two means.

$$\text{Standard error of difference} = \sqrt{\left(\frac{s_1^2}{n_1} + \frac{s_2^2}{n_2}\right)}$$

- s_1^2 is the standard deviation for data set 1.
- n_1 is the number of individual measurements in data set 1.
- s_2^2 is the standard deviation for data set 2.
- n_2 is the number of individual measurements in data set 2.

5. Compare the difference between the means with the standard error of the difference.

- If the difference between the means is greater than or equal to twice the standard error, you can be 95% confident that the results represent two different populations and can reject the null hypothesis.
- If the difference between the means is less than twice the standard error there is no significant difference and the null hypothesis has to be accepted. There is no difference between the samples at the $p = 0.05$ level. They may be part of the same set of data.

Note: Null hypothesis
There is no difference between the sets of data.

Note: When calculating the mean and standard deviation the central value in each class interval is used for all specimens in that class interval.

Example

The following data for lengths of worms were collected in two adjacent gardens. Test whether there is a significant difference between the two samples.

Class interval length of worm/mm	Garden 1	Garden 2
30–39	2	
40–49	5	
50–59	8	
60–69	14	2
70–79	17	6
80–89	14	9
90–99	10	14
100–109	7	18
110–119	3	12
120–129	1	7
130–139		5
140–149		3

For garden 1:

Number in sample, n_1 = 81
Mean, $\bar{x}_1$ = 76.9 mm
Standard deviation, s_1 = 19.5 mm

For garden 2:

Number in sample, n_2 = 76
Mean, $\bar{x}_2$ = 103.8 mm
Standard deviation, s_2 = 18.8 mm

Difference between the means = 26.9 mm

Standard error in difference =

$$\sqrt{\left(\frac{19.5^2}{81} + \frac{18.8^2}{76}\right)} = \sqrt{(4.69 + 4.65)} = 9.34 = 3.1 \text{ mm}$$

Twice the standard error in the means = 6.2 mm. If the difference between the means had been 6.2, we could have said that the populations were different with 95% confidence. Because the difference is over four times this (103.8 – 76.9 = 26.9) we can be even more confident that they are different populations.

Note: The statistics only tell you that you can be 95% confident that there is a difference. The statistics cannot tell you **why** there is a difference between the samples. You would need to do other tests to find this out.

Chi-squared (TEST A)

TEST A is used to test a set of data against a theoretical distribution or to compare two sets of data from the same experiment to see whether they are in agreement.

When is chi-squared test A used?

1. When you want to compare the data collected with data that would be expected using a particular model (theoretical data).
2. When you want to test whether the data are what would be expected if the data were purely random.
3. When you want to compare two sets of experimental data to see whether there is any difference between them.

Carrying out the test

1. Collect the data and make a table with rows and columns that indicate the characteristics being examined.
2. Calculate the values that would be expected in each cell of the table. You will then have two values for each characteristic: the actual observed value O and the value expected E.
3. Calculate the value of χ^2:
   - For each cell calculate the value of $(O-E)$, square this to find $(O-E)^2$ and then divide by E to find

     $$\frac{(O-E)^2}{E}$$

   - Add together all the values of $(O-E)^2/E$. This is χ^2 (chi-squared). Using mathematical notation,

     $$\chi^2 = \sum\frac{(O-E)^2}{E}$$

4. Determine the **degrees of freedom** in the data.

   The degrees of freedom for one column (or row) containing n items of data is $(n-1)$.

Drawing conclusions

If the calculated value of χ^2 is greater than or equal to the critical value, the observed data differ significantly from the expected data.

The larger the value of χ^2 the more confident we can be that the null hypothesis (that there is **no** difference) is incorrect.

If the calculated value of χ^2 is less than the critical value, the null hypothesis has to be accepted.

If the data contain only two categories, the value of χ^2 is given by

$$\chi^2 = \sum\frac{(|O-E|-0.5)^2}{E}$$

$|O-E|$ means the modulus of the difference. The difference is taken as a positive value whether O is bigger or smaller than E.

This is called the **Yates correction**.

Note: Null hypothesis
There is no difference between the two sets of data.

Critical values for different degrees of freedom

The critical values given here are at the 0.05 (or 5%) level. This means that if the calculated value of χ^2 is equal to the critical value the observed values had a 1 in 20 probability of occurring by chance. For higher values of χ^2 the chance is even lower.

For a more stringent test other tables provide critical values at the 0.01, 0.001 or even the 0.0001 level.

For example, for data with 5 degrees of freedom, for the null hypothesis to be disproved at the 0.0001 level, the value of χ^2 would have to exceed 25.74.

For your field work, the 5% level is usually considered adequate. For 5 degrees of freedom the critical value would be 11.08.

Degrees of freedom	Critical value at $p = 0.05$ level	Degrees of freedom	Critical value at $p = 0.05$ level
1	3.84	16	26.30
2	5.99	17	27.59
3	7.82	18	28.87
4	9.49	19	30.14
5	11.08	20	37.57
6	12.59	21	32.67
7	14.07	22	33.92
8	15.51	23	35.18
9	16.92	24	36.43
10	18.31	25	37.65
11	19.68	26	38.88
12	21.03	27	40.11
13	22.36	28	41.34
14	23.68	29	42.56
15	25.00	30	43.77

Chi-squared (TEST B) 2 × 2 contingency test

TEST B is used to test whether there is a relationship between two quantities using a **contingency table**. Only the test for a **2 x 2 contingency table** is considered here but the test can be extended to cover a wider range of variables.

When is chi-squared test B used?

1. When you have sample data for the corresponding frequencies of two variables.
2. When you want to test whether the data are what would be expected if the data were purely random or whether there is a relationship between the two variables.

Carrying out the test

1. Collect the data and make a table with rows and columns that indicate the occurrence of the characteristics being examined.
2. Calculate the values that would be expected in each cell of the table. The box to the right shows you how to do this. You will then have two values for each characteristic: the actual observed value O and the value expected E.
3. Calculate the value of χ^2.
   - **For each cell** calculate the value of $(O-E)$, square this to find $(O-E)^2$ and then divide by E to find

     $$\frac{(O-E)^2}{E}$$

   - Add together all the values of $(O-E)^2/E$. This is χ^2. Using mathematical notation,

     $$\chi^2 = \Sigma\frac{(O-E)^2}{E}$$

4. Calculate the number of **degrees of freedom**. If the table of data has R rows and C columns,

   Degrees of freedom = $(R - 1) \times (C - 1)$

5. Look up the critical value in the table (see page 77) that corresponds to the degrees of freedom.

Drawing conclusions

The larger the value of χ^2 the more confident we can be that the null hypothesis (that there is **no** difference) is incorrect.

If the calculated value of χ^2 is less than the critical value, the null hypothesis has to be accepted and there is no correlation between the factors.

Calculating the expected values for a 2 × 2 contingency table

In the table a, b, c and d are the observed values of two different factors (for example, a test of male and female colour blindness).

	Males	Females	Row total
Colour blind	a	c	$a + c$
Not colour blind	b	d	$b + d$
Column total	$a + b$	$c + d$	$a + b + c + d$

If there were no difference between males and females each should display the same proportion of colour blind individuals. The proportion of colour blind people should be

$$\frac{a+c}{a+b+c+d}$$

In the sample of males the expected number of colour blind males is

$$\frac{a+c}{a+b+c+d} \times \text{number of males} = \frac{(a+c)(a+b)}{a+b+c+d}$$

Similarly, the expected number of colour blind females is

$$\frac{(a+c)(c+d)}{a+b+c+d}$$

The expected value in each cell is therefore found by

$$\frac{\text{row total} \times \text{column total}}{\text{grand total}}$$

Example: Colour blindness

	Males	Females	Row total
Colour blind	28 (21)	7 (14)	35
Not colour blind	377 (384)	268 (261)	645
Column total	405	275	680

The figures in brackets are the expected values.

The null hypothesis is that there is no difference between the frequency of occurrence of colour blindness in males and females.

$$\chi^2 = 2.33 + 0.13 + 3.50 + 0.19 = 6.15$$

There is one degree of freedom, $(2 - 1) \times (2 - 1)$, since there are two rows and two columns of data. The critical value for $p = 0.05$ is 3.84. The calculated value is much greater than this, so on this evidence the null hypothesis is rejected.

The difference in occurrence of colour blindness in this sample is greater than would be expected to occur by chance. The data would therefore suggest that a higher proportion of males than females are colour blind.

Note: Null hypothesis
There is no difference between the observed frequencies and those expected by chance. There is no relationship between the variables.

USING THE CHI-SQUARED TEST

Example 1

Two students, A and B, collected data about the heights of male students:

	Height		
	More than 1.7 m	1.5 to 1.7 m	Less than 1.5 m
Number of A's students	15	35	10
Number of B's students	11	25	12

Is there a significant difference between the sets of data?

The null hypothesis is that there is no difference. The expected proportion of students in each category should be the same in both sets of data.

Using A's data we calculate the number of students that B would have expected in their sample. In A's sample, 15 of 60 students were more than 1.7 m tall. This is 0.25 of the students. B used a sample of 48 students and we would expect $0.25 \times 48 = 12$ students to be over 1.7 m tall.

Doing similar sums for the other heights gives the following table for B's observed and expected data.

	Height		
	More than 1.7 m	1.5 to 1.7 m	Less than 1.5 m
Expected number based on A's data	12	28	8
Number of students observed by B	11	25	12

χ^2 for this data = 0.08 + 0.32 + 2.00 = 2.40.

The data being examined are in B's single row so the degrees of freedom are (3 − 1) = 2. The table on page 77 gives the critical value as 5.99. As χ^2 is less than the critical value, the null hypothesis is correct and there is no significant difference between the sets of data.

Example 2

A genetic model suggests that if a pink flower self-pollinates the result should be red, pink and white flowers in the ratio 1 : 2 : 1.

The following data were collected in an experiment.

Red flowers	Pink flowers	White flowers
46	130	57

Are these data consistent with the theory?

The null hypothesis is that the theoretical and observed data agree.

The number of flowers in the observed sample is 233, and $\frac{1}{4}$ should be red, $\frac{1}{2}$ should be pink and $\frac{1}{4}$ should be white, so the expected data are

Red flowers	Pink flowers	White flowers
58	117	58

$\chi^2 = 2.48 + 1.44 + 0.02 = 3.94$.

There are three pieces of data in a single row so there are (3 − 1) = 2 degrees of freedom. The critical value is 5.99. As χ^2 is less than the critical value, the null hypothesis is accepted and the data are in line with the theoretical data.

Questions to try

1. The following data are from a test of a new drug for the treatment of angina. Some patients were treated with the drug and some with a harmless salt solution called a placebo. The table shows the frequency of the sample who were angina free and those who continued to suffer from angina in the two groups.

	Treated with new drug	Treated with placebo
Patients with angina	35	23
Patients without angina	110	97

Do the data suggest that the new drug is effective in the treatment of angina?

2. Five fishermen (all equally skilful) fished in five similar ponds for the same time. The number of fish that each caught in each pond was:

Pond	1	2	3	4	5
Number of fish	103	76	89	95	111

Do these data suggest that there is any significant difference between the number of fish in each pond?

3. In a genetics experiment using maize, the expected ratio for coloured smooth grain to colourless smooth grain to coloured wrinkled grain to colourless wrinkled grain is 9:3:3:1.

The measured outcome in the experiment was 134:36:49:14. Do the experimental data agree with the theoretical prediction?

If you want to know more about:
Probability see page 69
The chi-squared test see pages 77–78

The *t* test

As in the previous test this compares means. This test, however, can be used when you have fewer data. The test is sometimes called Student's *t* test. This is because when it was proposed the person who invented the test (W. S. Gossett) signed it 'Student' rather than using his own name. The value calculated in the test is the ratio of the difference between the means and the standard error in the difference.

THE *t* TEST FOR TWO INDEPENDENT SAMPLES

When is the test used?

1. When you are testing unrelated independent samples of data, e.g. from two separate experiments.
2. When you want to compare two sets of data, each of which has a normal distribution.
3. The more data you have the better, but the *t* test can be used with fewer data than other tests.
4. When used with smaller population sizes the test is less reliable, and this should be expressed in any conclusions drawn.

Carrying out the test

1. Check that both sets of data have a normal distribution. If you are measuring a length or weight this is almost certainly the case. If you are in doubt, draw a histogram to check that the distribution appears normal.
2. Calculate the mean for each set of data, $\bar{x}_1$ and $\bar{x}_2$:

   $$\text{Mean} = \bar{x} = \frac{\Sigma x}{n}$$

3. Calculate the magnitude of the difference between the two means. This is taken as positive.

   $$\text{Difference between the means} = |\bar{x}_1 - \bar{x}_2|$$

4. Calculate the value of *t*:

   $$t = \frac{\text{difference between the means}}{\text{standard error in the difference}}$$

5. Calculate the number of **degrees of freedom** – the total population in both samples minus 2:

   $$\text{Degrees of freedom} = (n_1 + n_2 - 2)$$

   If you have only one column of data the number of degrees of freedom is (R – 1), where R is the number of data items in the column.

6. Look up the critical value that corresponds to the degrees of freedom in the table that gives critical values of *t*. The table is on page 81.

Drawing conclusions

If the calculated value of *t* is greater than or equal to the critical value, the observed data differ significantly from the expected data.

The greater the value of *t* the more confident we can be that the null hypothesis is incorrect.

If the calculated value of *t* is less than the critical value, the null hypothesis has to be accepted.

Note: Null hypothesis

There is no difference between the two sets of data.

How do you calculate the standard error in the difference?

Calculate the square of the standard deviation s^2 for each set of data:

$$s^2 = \frac{\Sigma x^2}{n} - \bar{x}^2$$

Note that, for small samples, $n-1$ should be used and the square of the standard deviation of the sample population is given by

$$s^2 = \frac{\Sigma(x - \bar{x})^2}{n-1}$$

Calculate the standard error in the difference between the means.

$$\text{Standard error in difference} = \sqrt{\left(\frac{s_1^2}{n_1} + \frac{s_2^2}{n_2}\right)}$$

i.e. divide the value of s^2 for each set of data by the number of observations in the sample, add these together and take the square root.

Example: Unpaired *t* test

Equal numbers of seeds were sown in two different composts. For compost A seven trays were prepared, and for B five trays. The number of seeds that germinated was counted for each tray. Is there a significant difference between the germination rates in the two composts?

	Tray						
	1	2	3	4	5	6	7
Compost A	20	25	30	15	28	27	38
Compost B	25	38	44	27	21		

The null hypothesis is that there is no difference in the germination rates.

- Mean for compost A = 26.1, s = 7.4 (using $n-1$).
- Mean for compost B = 31.0, s = 9.6.
- Difference between the means = 4.9.
- Standard error in difference $= \sqrt{\left(\frac{7.4^2}{7} + \frac{9.6^2}{5}\right)} = 5.1$.
- $t = 4.9/5.1 = 0.96$.
- Degrees of freedom = 7 + 5 – 2 = 10.
- Critical value = 2.23 at p = 0.05 confidence level. The *t* value is much less than the critical value, so there is no significant difference between the germination rates in the two composts.

If you want to know more about:

Means see page 73

Standard deviation see page 74

Using a calculator to calculate statistical data see page 75

When is the test used?

1. When you are testing two samples of data that are linked in some way.
2. When you want to compare two sets of data, each of which has a normal distribution.
3. The more data you have the better, but this test can be used with fewer data than other tests.
4. When used with smaller population sizes the test is less reliable, and this should be expressed in any conclusions drawn.

For example, the test could be used to compare the frequency of a species of plant on the north and south sides of a hedge, or how two features vary for samples of two groups of individuals who have been matched in some way, such as differences in characteristics of male–female twins.

Carrying out the test

1. You will have *n* items in each set of data. Check that both sets of data have a normal distribution. If you are in doubt, draw a histogram to check that the distributions appear normal.
2. Calculate the differences (*d*) between each pair of data in your sample.
3. Calculate the **mean of the differences**, $\frac{(\Sigma d)}{n}$. This turns out to be the same as the difference between the two means.
4. Calculate the **standard deviation of the differences**:

$$s = \sqrt{\left(\frac{(d-\bar{d})^2}{n-1}\right)}$$

5. Calculate the standard error in the difference between the means:

$$\text{SE} = \frac{\text{standard deviation}}{\text{square root of number of pairs of measurements}} = \frac{s}{\sqrt{n}}$$

6. Calculate the value of *t*:

$$t = \frac{\text{mean of the differences}}{\text{standard error in the differences}}$$

7. Calculate the number of **degrees of freedom**.

   Degrees of freedom = $(n-1)$

   i.e. the number of pairs of data minus 1.
8. Look up the critical value that corresponds to the degrees of freedom in the table that gives critical values of *t*.

Note: Null hypothesis
There is no difference between the two sets of data.

Question to try

The table shows the moisture content at two different heights along the stem for a number of plants. Is there a significant difference between the data at the two heights?

	Plant					
	1	2	3	4	5	6
h_1	51.5	55.3	52.7	47.3	58.6	55.2
h_2	45.4	52.1	46.9	40.5	54.0	50.8

Critical values of *t*

Degrees of freedom	Critical value	
	$p = 0.05$	$p = 0.01$
5	2.57	4.03
10	2.23	3.17
12	2.18	3.06
14	2.15	2.98
16	2.12	2.92
18	2.10	2.88
20	2.09	2.85
21	2.08	2.83
22	2.07	2.82
23	2.07	2.81
24	2.06	2.80
25	2.06	2.79
26	2.06	2.78
27	2.05	2.77
28	2.05	2.76
29	2.05	2.76
30	2.04	2.75
40	2.02	2.70
60	2.00	2.66
100	1.98	2.62
∞	1.96	2.58

These critical values are those at the 0.05 (or 5%) level and the 0.01 (or 1%) level. At the 0.01 level, if the value of *t* is equal to the critical value the observed values had a 1 in 100 probability of occurring by chance. At the 0.01 level if *t* is greater than the critical value you are 99% confident that the two distributions are for different populations.

For your field work, the 5% level will usually be adequate.

Drawing conclusions

If the calculated value of *t* is greater than or equal to the critical value, the two sets of data show significant differences. The greater the value of *t* the more confident we can be that the null hypothesis is incorrect.

Example: A paired *t* test

To test whether a particular fertilizer improved the yield of tomatoes, six pairs of plants were used. Fertilizer was used with one of each pair chosen at random. After a set time the tomato yield was weighed. The table gives the masses of tomatoes in kg. Do the data suggest that the fertilizer is effective?

	Pair					
	1	2	3	4	5	6
No fertilizer	3.2	3.7	2.6	3.5	4.0	3.0
Fertilizer	4.0	3.6	3.5	4.4	3.9	3.3

Differences between pairs: 0.8, –0.1, 0.9, 0.9, –0.1, 0.3.

Mean of differences = 0.52.

Standard deviation = 0.39.

Standard error = $\frac{0.39}{\sqrt{6}}$ = 0.16.

$t = \frac{0.52}{0.16} = 3.25$.

Degrees of freedom = 6 – 1 = 5.

Critical value = 2.57.

Because *t* is greater than the critical value at $p = 0.05$ level we can be more than 95% confident that the fertilizer improves the yield.

Mann–Whitney *U* test

When is the Mann–Whitney *U* test used?

1. When you want to compare two sets of data which do not have normal distributions. The data may show no particular pattern. The histogram may show a skewed distribution.
2. When you do not have a large sample size but have at least six values in each set of data.

Note: Mann–Whitney or t test?
If you have more than 20 in the sample it is probably best to use the *t* test even though the data may not be a true normal distribution.

Carrying out the test

You will have n_1 values of data in set 1 and n_2 values of data in set 2.

1. Arrange **all** the data in rank order, starting with the lowest value.
2. Allocate a number corresponding to the rank order position to each data value.
3. Where there is a 'tie' of two or more pieces of data, allocate the mean of the positions occupied to each one. For example:
   - If there are two equal values in positions 3 and 4 they would both be allocated rank score 3.5.
   - If there are three equal values in positions 5, 6 and 7, they would be allocated rank score 6.
   - If there are four equal values in positions 8, 9, 10 and 11 they each have the rank score 9.5.
4. Add together the rank scores for each set of data. For two sets of data, 1 and 2, you now have

   ΣR_1 and ΣR_2

5. Calculate the *U* score for each set of data.

   $$U_1 = n_1 n_2 + \frac{n_2(n_2 + 1)}{2} - \Sigma R_2$$

   $$U_2 = n_1 n_2 + \frac{n_1(n_1 + 1)}{2} - \Sigma R_1$$

   If you have not made any mistakes

   $U_1 + U_2 = n_1 n_2$

   If this is not true, check your working.

6. Select the **smaller** of the two *U* values.
7. Look up the critical value that corresponds to the values of n_1 and n_2 from the table.

Drawing conclusions

If the **smaller** *U* value is **equal to or less than** the critical value, there is a significant difference between the two sets of data.

The smaller the *U* value the more confident you can be that the null hypothesis is incorrect.

Note: Using a spreadsheet
This process can be done by hand, but is more easily managed using a spreadsheet.

- Enter each set of data in a column.
- Use the spreadsheet to place each column into rank order. (Leave space for inputting the rank order later.)
- Copy both sets of data into another column so that the second set follows on from the first set.
- Use the spreadsheet to place this in ascending rank order.
- Allocate the rank order for each piece of data starting with 1.
- Where there is more than one value give all pieces of data the average for the rank position occupied.
- Enter the correct rank order values in the columns alongside the data for each set.

The spreadsheet can now be used to total the rank scores and calculate the *U* values.

Note: Null hypothesis
There is no difference between the two sets of data.

Critical values of *U*

Number of data values in set 1 ↓ / Number of data values in set 2 →	1	2	3	4	5	6	7	8	9	10	11	12	13	14	15	16	17	18	19	20
1																				
2								0	0	0	0	1	1	1	1	1	2	2	2	2
3					0	1	1	2	2	3	3	4	4	5	5	6	6	7	7	8
4				0	1	2	3	4	4	5	6	7	8	9	10	11	11	12	13	13
5			0	1	2	3	5	6	7	8	9	11	12	13	14	15	17	18	19	20
6			1	2	3	5	6	8	10	11	13	14	16	17	19	21	22	24	25	27
7			1	3	5	6	8	10	12	14	16	18	20	22	24	26	28	30	32	34
8		0	2	4	6	8	10	13	15	17	19	22	24	26	29	31	34	36	38	41
9		0	2	4	7	10	12	15	17	20	23	26	28	31	34	37	39	42	45	48
10		0	3	5	8	11	14	17	20	23	26	29	33	36	39	42	45	48	52	55
11		0	3	6	9	13	16	19	23	26	30	33	37	40	44	47	51	55	58	62
12		1	4	7	11	14	18	22	26	29	33	37	41	45	49	53	57	61	65	69
13		1	4	8	12	16	20	24	28	33	37	41	45	50	54	59	63	67	72	76
14		1	5	9	13	17	22	26	31	36	40	45	50	55	59	64	67	74	78	83
15		1	5	10	14	19	24	29	34	39	44	49	54	59	64	70	75	80	85	90
16		1	6	11	15	21	26	31	37	42	47	53	59	64	70	75	81	86	92	98
17		2	6	11	17	22	28	34	39	45	51	57	63	67	75	81	87	93	99	105
18		2	7	12	18	24	30	36	42	48	55	61	67	74	80	86	93	99	106	112
19		2	7	13	19	25	32	38	45	52	58	65	72	78	85	92	99	106	113	119
20		2	8	13	20	27	34	41	48	55	62	69	76	83	90	98	105	112	119	127

USING THE MANN–WHITNEY *U* TEST

Example

A fruit grower has a number of apple trees of two different varieties. The grower wants to assess whether one is a better producer of fruit than the other. In one season the fruit collected from each tree was weighed and the following data tabulated.

Tree variety	Mass of fruit/ kg															
A	30	34	25	33	35	21	36	44	36	38	24	35	41	36	31	29
B	34	22	27	43	39	28	36	24	27	37	36	23				

The null hypothesis is that there is no difference between the produce from the two varieties of tree.

First we place the data in rank order, obtain the rank score for each item and then the sum of the rank scores for each set of data.

Data A	Rank score	Data B	Rank score	All data	Rank score
21	1	22	2	21	1
24	4.5	23	3	22	2
25	6	24	4.5	23	3
29	10	27	7.5	24	4.5
30	11	27	7.5	24	4.5
31	12	28	9	25	6
33	13	34	14.5	27	7.5
34	14.5	36	19	27	7.5
35	16.5	36	19	28	9
35	16.5	37	23	29	10
36	19	39	25	30	11
36	19	43	27	31	12
36	19		161	33	13
38	24			34	14.5
41	26			34	14.5
44	28			35	16.5
	240			35	16.5
				36	19
				36	19
				36	19
				36	19
				36	19
				37	23
				38	24
				39	25
				41	26
				43	27
				44	28

Note that there are five values of 36. The lowest is in position 17 and the highest in position 21. All five therefore take the rank score 19, the mean of the positions occupied.

This is ΣR for tree variety A

This is ΣR for tree variety B

Calculating the *U* values

The *U* value is calculated for each set of data.

For tree variety A,

$$U = (16 \times 12) + 0.5(12 \times 13) - 161 = 109$$

For tree variety B,

$$U = (16 \times 12) + 0.5(16 \times 17) - 240 = 88$$

The table gives a critical value of 53 for 16 values in one set of data and 12 values in the other. The lower of the *U* values is greater than the critical value. **There is therefore no significant difference between the produce of the two varieties of tree.**

Question to try

The cotton on two reels looks similar. Two students perform experiments, each using a different reel, to determine the force that breaks a number of samples. The data for each student are given below. Test whether the students' data suggest that there is any difference between the cotton on each reel.

Student	Breaking strength/N					
A	10.5	11.5	12.6	12.1	10.7	
B	11.3	12.8	11.8	11.8	12.8	11.0

Which test?

In an investigation you will sometimes have two sets of data that you want to compare. It is important that you choose the right test for the comparison. If you use the wrong test, your findings will be meaningless.

TESTING WHETHER SETS OF DATA ARE SIGNIFICANTLY DIFFERENT

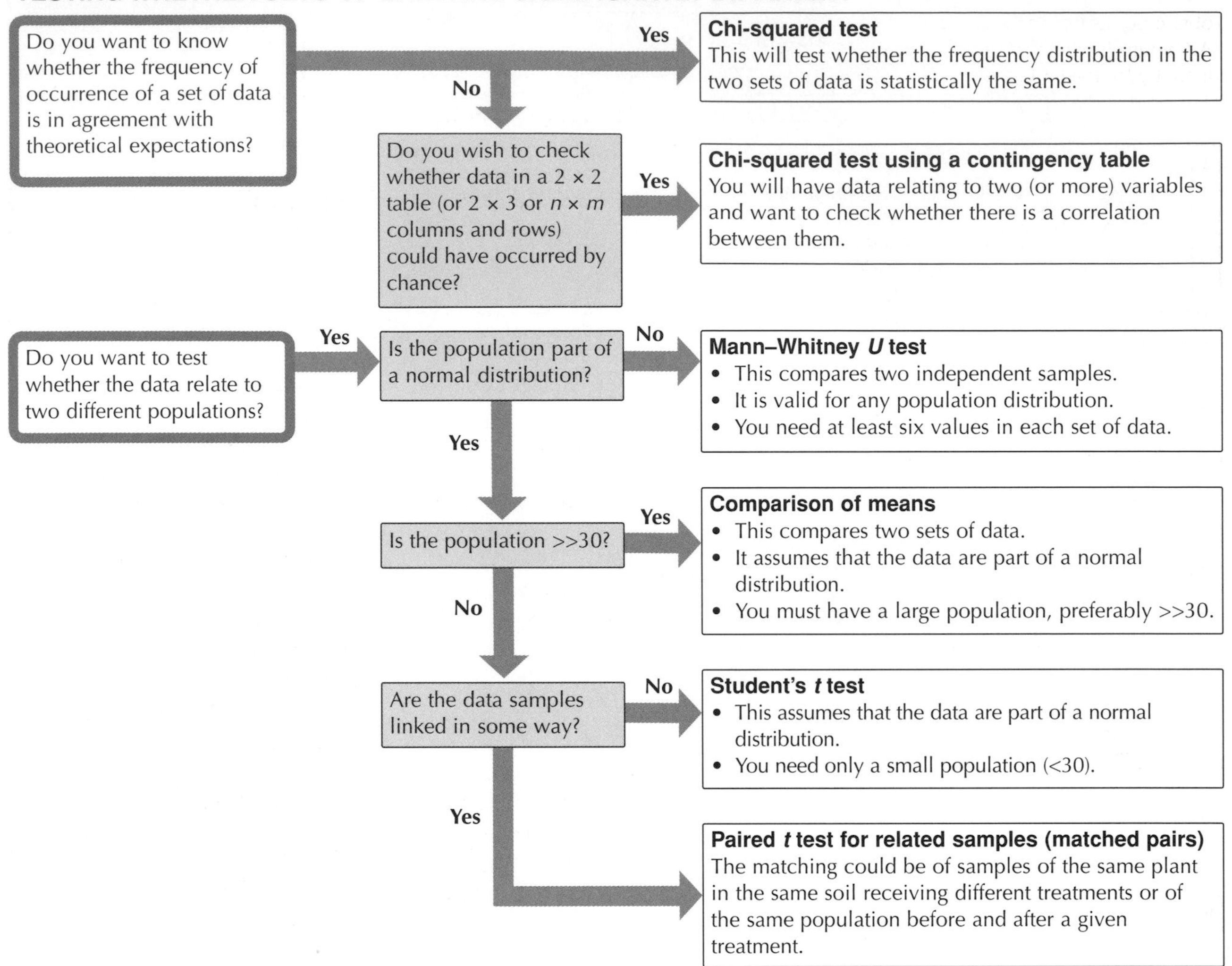

CORRELATION TESTS

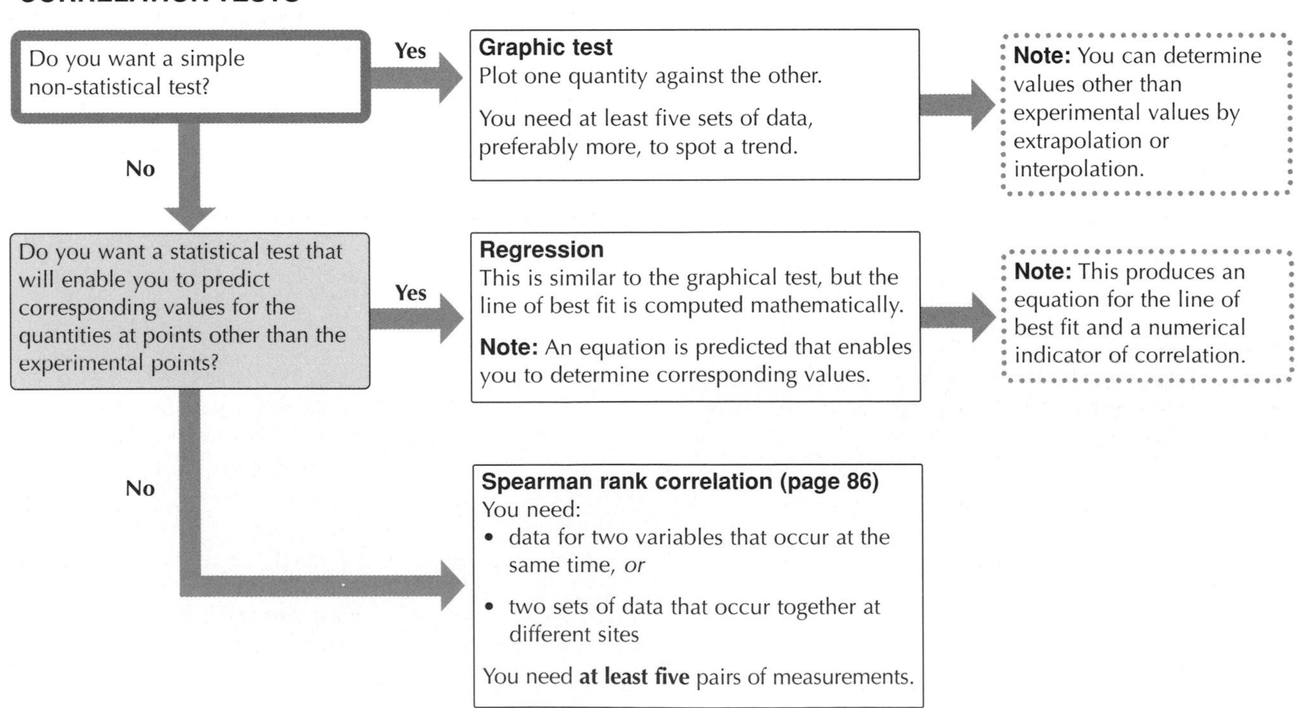

Correlation testing

A correlation test is simply a test to see whether there is any relationship between two quantities. This can be done by graphical representation of the data or by statistical analysis. When you plot data for two quantities on a graph you may obtain points like the ones shown in Graphs A–D (below). Ask yourself whether the points appear to represent a trend. If so, try to draw a line of best fit to show that trend.

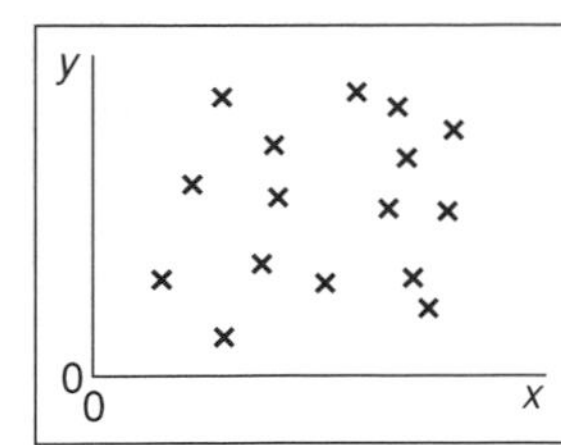

Graph A
This shows a random scatter of experimental points. No apparent trend is shown by the data and you would suggest no relationship between the quantities.

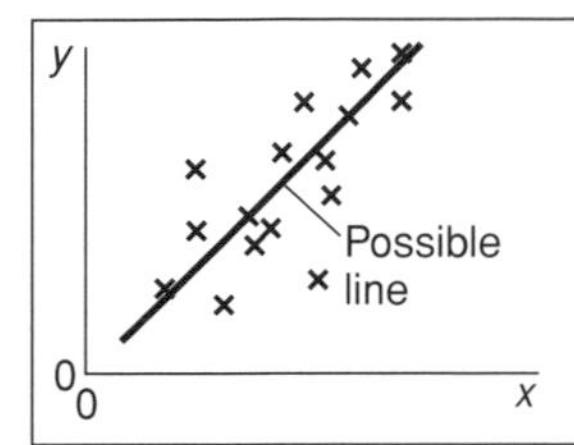

Graph B
As one quantity increases the other appears to increase. Although some points do not follow this trend, the data as a whole suggest a positive correlation.

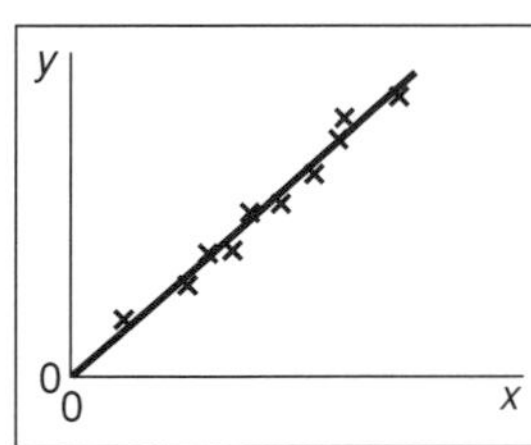

Graph C
Graph C shows a clear positive correlation, shown by the positive gradient for the line of best fit.

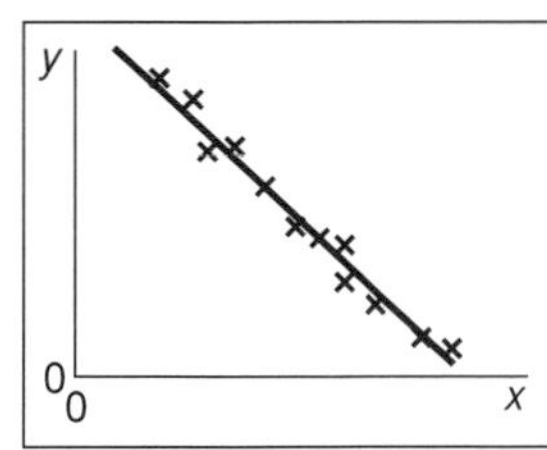

Graph D
Graph D shows a clear negative correlation. The line of best fit has negative gradient.

Note: The **line of best fit** is one that gives an even distribution of points about the line. This is not easy to place when there is poor correlation as in graph B. A simple test of the degree of correlation is used to draw the line of best fit easily.

Regression

For most of your work, a graphical test for correlation will be adequate. Sometimes, however, the trend is not very clear and it is not possible to draw a line of best fit. In this case you can undertake statistical analysis of the data. This will enable you to determine the equation for the line of best fit and obtain a statistic that tells you how closely the data fit the line. The process is long and complex by hand so here we will look at how you can use a scientific calculator that has a **linear regression** mode.

What do you calculate?

The equation for a straight line is

$$y = mx + c$$

The regression analysis determines m and c. You will then have the equation that relates the quantities that you have called x and y.

Determining the gradient *m* and intercept *c*

The process here is for one type of calculator. Check your handbook to see if your calculator is the same.

- Switch to linear regression mode: enter MODE then 2.
- Clear the memory: enter INV then AC.

Now enter the data.

- Enter the first x value; enter x_D, y_D.
- Enter the first y value; enter DATA.
- Repeat for each pair of values.
- To read out the value for c, press INV then A (the 7 key).
- To read out the value for m, press INV then B (the 8 key).

Determining the correlation coefficient

The **correlation coefficient** r tells us how well the data fit the line. The value will be in the range –1 to +1.

- –1 means perfect negative correlation.
- 0 means no correlation at all.
- +1 means perfect positive correlation.

It is found by pressing INV then r (the 9 key).

Determining points on your regression line

- To find a y value corresponding to an x value: insert the x value then K_{OUT} then $\bar{y}$.
- To find an x value corresponding to a y value: insert the y value then INV then $\bar{x}$.

If you want to know more about:
Straight-line graphs see pages 42–43 

Questions to try

Use the following data to test whether there is a correlation between molar heat of vaporization and the boiling point of the substances.

	Boiling point/ K	Molar heat of vaporization/ kJ mol^{-1}
Methane	112	8.2
Benzene	353	31
Sodium	1163	89
Lithium	1604	135
Sodium chloride	1740	170

1. (a) Plot the points on a graph.
   (b) Try to draw the best line through your points.
   (c) Give an estimate for the molar heat capacity of a substance with a boiling point of 700 K.
2. (a) Use a calculator to determine the gradient m and intercept c of the regression line.
   (b) What is the value of the correlation coefficient r? Does this figure suggest a good correlation?
   (c) What value does the regression line give for the molar latent heat of the substance in 1(c)?

Spearman rank correlation test

When is the Spearman rank correlation test used?

1. When you want to test whether there is a relationship between two measurements.
2. When you have at least five pairs of measurements for the two quantities that you are comparing.

Note: Null hypothesis
There is no relationship between the two quantities you are comparing.

Carrying out the test

1. Arrange in a table the corresponding data for the variables A and B that you are comparing (columns 1 and 3). You should have n pairs of measurements where $n \geq 5$.
2. Give the measurements in each column a rank score. Put these in columns 2 and 4. Where there is a tie between two or more measurements allocate the mean of the positions occupied to each one. For example:
   - If there are two equal values in positions 3 and 4 they would both be allocated rank score 3.5.
   - If there are three equal values in positions 5, 6 and 7 they would be allocated rank score 6.
   - If there are four equal values in positions 8, 9, 10 and 11 they would each have the rank score 9.5.

   These are the values R_A and R_B.
3. Calculate the difference d between each pair of rank scores. (Subtract column 4 from column 2 and insert the answer in column 5.)

   $d = (R_A - R_B)$
4. Calculate the value of d^2 for each pair. Place these values in column 6.
5. Add together the d values in column 5. The sum of the d values Σd should be 0. If it is not, check your rank scores.
6. Add together the d^2 values in column 6. This gives Σd^2.
7. Calculate the Spearman rank correlation coefficient r_S

   $$r_S = 1 - \frac{6\Sigma d^2}{(n^3 - n)}$$

   where n is the number of pairs. The value you obtain for the Spearman rank correlation coefficient should be between –1 and +1.
8. Look up the critical value that corresponds to the number of pairs of measurements n in the table.

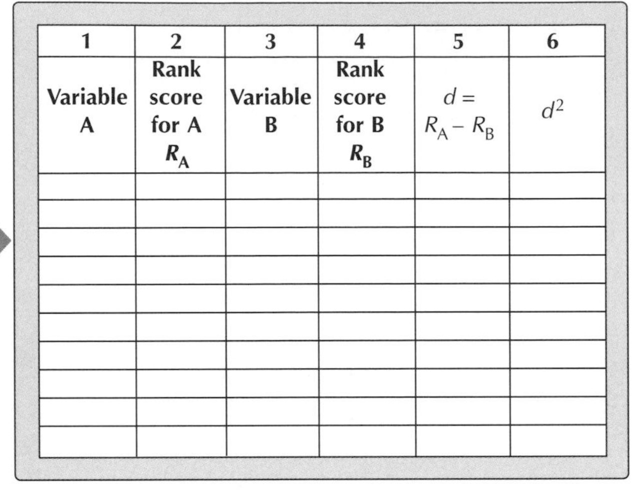

1 Variable A	2 Rank score for A R_A	3 Variable B	4 Rank score for B R_B	5 $d = R_A - R_B$	6 d^2

Critical values for r_s

Number of pairs of measurements	Critical value
5	1.00
6	0.89
7	0.79
8	0.74
9	0.68
10	0.65
12	0.59
14	0.54
16	0.51
18	0.48
20	0.45
22	0.43
24	0.41
26	0.39
28	0.38
30	0.36

Drawing conclusions

Ignore the sign: if the magnitude of r_S is **greater than** the critical value the null hypothesis is incorrect and there **is** a correlation between the two measurements or quantities.

If r_S is positive, there is a positive correlation. This means that the one quantity increases linearly as the other increases.

If r_S is negative, there is a negative correlation. This means that the one quantity decreases linearly as the other increases.

Note: Beware
Do not confuse the fact that there is correlation with cause and effect. Correlation does not mean that the change in one factor **causes** the other to change. They may both be changing because some other factor is influencing both of them.

Example

The following data relate the number of eggs produced by an insect and the body mass of the insect.

Body mass $/10^{-3}$ kg	Number of eggs produced
2.8	30
3.0	10
3.5	32
3.0	44
3.9	35
3.6	58

Test whether the data suggest that there is a correlation between the body mass of the insect and the number of eggs produced.

The null hypothesis is that there is no correlation.

1	2	3	4	5	6
Variable A	Rank score for A R_A	Variable B	Rank score for B R_B	$d = R_A - R_B$	d^2
2.8	1	30	2	−1.0	1.0
3.0	2.5	10	1	+1.5	2.3
3.5	4	32	3	+1.0	1.0
3.0	2.5	44	5	−2.5	6.3
3.9	6	35	4	+2.0	4.0
3.6	5	58	6	−1.0	1.0
				0	15.5
				Σd	Σd^2

For variable A the sequence of data is

2.8 3.0 3.0 3.5 3.6 3.9

There are two 3.0 values so they each have the score

$$\frac{(2+3)}{2} = 2.5$$

The Spearman rank coefficient

$$r_S = 1 - \frac{6 \times 15.5}{6^3 - 6}$$

$$= 1 - \frac{93}{210}$$

$$= 1 - 0.44 = 0.56$$

There are six pairs of measurements so the critical value is 0.89.

r_S is less than the critical value so the null hypothesis is correct. The data suggest that there is no correlation between the body mass of the insect and the number of eggs produced.

Questions to try

1. A student decided to investigate whether planting separation affects the height to which plants grow. The mean height of the plants in the test areas was calculated and the following data were obtained:

Planting separation/ cm	Mean height/ cm
5.0	24.5
6.0	21.5
7.0	27.3
8.0	30.0
9.0	26.0
10.0	29.2
11.0	24.5
12.0	29.0
13.0	32.0

Do these data suggest any correlation between the planting separation and the mean height of the plants?

2. The following data were obtained from a data book relating the melting point of some metallic solids and their specific latent heat of fusion.

Metal	Melting point/ K	Specific latent heat of fusion/ 10^4 J kg^{-1}
Aluminium	932	38
Aluminium alloy	800	39
Antimony	904	16
Bismuth	544	5
Cobalt	1765	25
Copper	1356	21
Iron	1810	27
Magnesium	924	38

Use the Spearman rank correlation test to determine whether the data suggest any correlation between melting point and the specific latent heat of fusion.

Simpson's diversity index

Diversity tells you about the richness of a particular site in terms of the different species of plant or other organisms present. The length of a list of different species is one measure of diversity, but it does not take account of the abundance of the different species present.

When is Simpson's diversity index used?

1. When you want to compare the diversity of one site with that of another similar site.
2. When you want to compare the diversity of a given site over a period of time.

Why is Simpson's diversity index used?

Simpson's diversity index is used to place a numerical value on the richness of species in a given site. It takes account of:

- the total number of species present
- the total number of individuals of each species present.

Determining the index

1. Sample the site using randomly placed quadrats. The more areas sampled the firmer your evidence will be.
2. In each quadrat list the different species present. You may be studying animals, insects, plants, etc.
3. Count the number of each species in each quadrat and total them in a table. This is the value n.
4.
   - For each species (each row) calculate the value of $n(n-1)$.
   - Add together all the values of $n(n-1)$. This gives $\Sigma n(n-1)$.
   - Calculate the total number N of all species present. This is the sum of all the n values:

     $N = \Sigma n$

   - Calculate the value of $N(N-1)$.
5. Calculate **Simpson's diversity index**:

$$D = \frac{N(N-1)}{\Sigma n(n-1)}$$

Species	Number found n	$n(n-1)$
1		
2		
etc.		
	$N = \Sigma n$	$\Sigma n(n-1)$

High diversity

Areas showing a high level of diversity would be those that have been allowed to develop naturally over a long period of time. They would not have been influenced by cultivation of land or been subject to severe natural disasters or human environmental influences.

Extreme conditions tend to reduce the number of different species that can coexist.

One objective of any investigation of diversity will be to suggest valid reasons for any changes in the diversity.

A **quadrat** is a clearly defined area in which data are collected. The region may be defined by a wooden frame of appropriate dimensions.

Example

A meadow was sampled for plants and the following data were obtained.

Species	Number found, *n*
Cowslip	42
Dandelion	23
Buttercup	35
Plantain	25
Lungwort	15

Calculate the Simpson diversity index for this meadow.

First complete the table with $n(n-1)$ values:

Species	Number found, *n*	$n(n-1)$
Cowslip	42	42 × 41 = 1722
Dandelion	23	23 × 22 = 506
Buttercup	35	35 × 34 = 1190
Plantain	25	25 × 24 = 600
Lungwort	15	15 × 14 = 210

$\Sigma n(n-1) = 1722 + 506 + 1190 + 600 + 210 = 4228$

$N = \Sigma n = 42 + 23 + 35 + 25 + 15 = 140$

$N(N-1) = 140 \times 139 = 19\,460$

$$D = \frac{N(N-1)}{\Sigma n(n-1)} = \frac{19\,460}{4228} = 4.60$$

Species frequency

This is the probability of finding a particular species in a random quadrat.

The frequency is equal to the number of quadrats that contain the particular species divided by the total number of quadrats in the sample.

To have real significance there must be a large number of quadrats in the test area. The size of the quadrats also has to be suitable for the size and general distribution of the plants.

When determining the frequency you need to decide the rules for presence of the species. You may decide that a plant species is present if any foliage overlaps the area, or you may decide to record its presence only if it is rooted in the particular quadrat. The rules you apply should be clearly stated in your findings.

The species frequency is usually expressed as a percentage:

$$\text{Presence} = \frac{\text{no. of quadrats species found in}}{\text{number of quadrats examined}} \times 100\%$$

Species abundance

This is found by having a quadrat frame that is divided into 25 squares.

The number of squares n in which the species is found is determined.

$$\text{Percentage abundance} = \frac{n}{25} \times 100\% = 4n$$

The percentage abundance is the number of squares multiplied by 4.

Question to try

A second meadow yielded the following data:

Species	Number found, *n*
Cowslip	20
Dandelion	41
Buttercup	15
Plantain	19
Lungwort	6
Daisy	12
Thistle	9

Does this meadow show a greater or smaller diversity than the example on the left?

If you want to know more about:
Mathematical symbols see page 7
Percentages see page 18

Population sampling

THE CAPTURE–RECAPTURE METHOD

How can we know how many fish there are in a lake without catching them all? How can you count how many wood pigeons there are when they keep flying around? It is, of course, impossible to count them individually but you can still obtain a good estimate of the population by a process known as the **capture–recapture** method. As with all statistical methods, the greater the sample the better the reliability of the estimate.

What is the capture–recapture method?

1. This technique involves capturing a random sample of the species. The sample is counted and each individual is marked in some way. Metal or plastic tags or a marker dye or paint could be used, for example.

   Total number marked = N

2. The sample is now released into the area from which it was taken.
3. Time is allowed for the sample to mix randomly with the rest of the population.
4. A new random sample is collected. This does not have to be the same size as the first sample.
   - The total number in the sample, M, is counted.
   - The number of those captured that were previously marked, m, is also counted.
5. The estimated number in the population is called the **Lincoln index**.

$$\text{Estimated total population} = \frac{N \times M}{m}$$

$$= \frac{\text{total number marked} \times \text{total recaptured}}{\text{number recaptured that were marked}}$$

When can the capture–recapture method be used?

1. The population must exist in a fixed area. There must be limits outside which the individuals cannot go.
2. There must be no changes in the population of the species due to individuals joining from outside.
3. The time scale of the investigation must be short enough for births and deaths of the species to be negligible compared with the total population.
4. The individuals must mix randomly within the area. This means that you cannot use it when the species roams in groups (as in herds of bison or shoals of fish).

What conditions are needed for a good result?

The investigation relies on statistics so that randomness is important.

1. The sample must be selected randomly before and after release.
2. Time must be allowed for the captured sample to mix randomly again with the rest of the population.
3. The observation must not influence the sample by introducing bias. Marking of the sample must not restrict movement so that the individuals do not mix randomly or make the individuals more likely to be killed by predators.

Assumption of the capture–recapture method

The method assumes that the proportion of marked species in the second sample is the same as the proportion of the total number marked in the whole population. The conditions described make this assumption reasonable. This means that

$$\frac{\text{number in the first sample}}{\text{total population}} = \frac{\text{number marked in the recaptured sample}}{\text{total in the second sample}}$$

Example

To estimate the number of fish in a lake, 350 were caught and tagged. The fish were released. A few days later 250 fish were caught and of these 74 were marked. Estimate the number of fish in the lake.

$$\text{Total number of fish} = \frac{\text{first sample size} \times \text{second sample size}}{\text{number marked}}$$

$$\text{Estimated number of fish} = \frac{350 \times 250}{74} = 1180$$

Question to try

A local council wanted to determine the pigeon population in the city centre, so 50 pigeons were caught and rings were placed around their legs. Two days later 40 birds were captured and of these seven had rings on their legs.

(a) What is the estimated population of pigeons in the city centre?

These birds were all released and a week later 50 birds were captured of which five were found to have rings.

(b) What would be the new estimate of the pigeon population?

(c) Consider the conditions necessary for the capture–recapture method to work effectively and discuss possible causes of the difference between the two estimates.

CAPTURE WITH PARTIAL RELEASE METHOD

Another problem that may arise is to determine, for example, the total population and the number of males and females in a population. You could do this by the capture–recapture method, but sometimes releasing of all those captured is not the best option. The method described here involves capturing a random sample and removing all or some of the males or females before releasing the remainder.

For the method to work the sample must be large enough to be representative of the population as a whole.

Principle of the method

Suppose that the first sample contains *a* males and *b* females of a species. The ratio of males to females (M:F) in the captured population equals the ratio of the total number of males N_1 to the total number of females N_2 in the whole population.

$$\frac{a}{b} = \frac{N_1}{N_2} \quad \text{Equation A}$$

1. Some males or females are removed from the sample and retained in captivity. Suppose *n* males are removed.
2. The sample is now released back into the community. The number of males in the community is now $N_1 - n$.
3. Time is allowed for the sample to mix randomly with the rest of the population.
4. A random sample is now collected. (This need not be the same size as the first.) The male and female populations in the recaptured sample are counted.
5. Suppose that there are *c* males and *d* females. Provided the sample is large enough this is now representative of the new population (the old population less the males retained in captivity).

$$\frac{c}{d} = \frac{N_1 - n}{N_2}$$

$$\frac{c}{d} = \frac{N_1}{N_2} - \frac{n}{N_2} \quad \text{Equation B}$$

6. A and B are simultaneous equations. Subtracting B from A,

$$\left(\frac{a}{b} - \frac{c}{d}\right) = \frac{n}{N_2}$$

$$N_2 = \frac{n}{\left(\frac{a}{b} - \frac{c}{d}\right)}$$

This gives the number of females in the original population.

Number of females

$$= \frac{\text{number of males removed from population}}{\text{difference in M:F ratios in the two samples}}$$

7. Substituting in A gives the number of males in the original population:

$$N_1 = \frac{a}{b} N_2 = \frac{a}{b}\left(\frac{n}{\frac{a}{b} - \frac{c}{d}}\right)$$

$$= \frac{\text{number removed} \times \text{M:F ratio in first sample}}{\text{difference in M:F ratios in the two samples}}$$

8. The total population of males plus females is $N_1 + N_2$.

$$\text{Total population} = \left(\frac{n}{\frac{a}{b} - \frac{c}{d}}\right)\left(1 + \frac{a}{b}\right)$$

$$= \frac{\text{number removed} \times (1 + \text{original M:F ratio})}{\text{difference in M:F ratios in the two samples}}$$

Note: Conditions required

The same conditions apply as for the recapture method on page 90. You need to be considering a closed system and using a large random sample. There can be no immigration or emigration, births and deaths must be negligible, and there must be random mixing of the population.

Example

A pest exterminator wants to know the mouse population in a given environment. Having caught the mice, as few as possible are to be released back into the environment; so the males are exterminated. The first sample produces the following data.

Number of males = 30

Number of females = 50

After some days a new sample yields the following data:

Number of males = 25

Number of females = 60

Estimate the number of males and females in the new population.

Original M:F ratio = 30/50 = 0.60
Final M:F ratio = 25/60 = 0.42
Difference in ratios = 0.18
Female population = number removed/0.18
= 30/0.18 = 167

Final male population = (25/60) × female population
= 70

The original population would have been 100 males and 167 females (a ratio of 0.6 as in the first sample).

Question to try

There are known to be two species of fish in a lake. A sample of 100 was collected of which $\frac{1}{4}$ were of species A. Only fish of species B were returned to the lake. Some time later a further sample was collected of which $\frac{1}{5}$ was species A. Estimate the number of fish of both species that are in the lake. How many were of species A when the first sample was collected?

If you want to know more about:
Ratios see page 17
Simultaneous equations see pages 22–23

Formulae and relationships between physical quantities

In AS and advanced-level physics examinations you will not be given the following formulae. For advanced level you need to be able to recall and use all of them. For AS the equations you need to recall and use will depend on the syllabus you are following. You will usually be given a formula sheet that includes other equations relevant to your syllabus.

Speed = $\frac{\text{distance}}{\text{time}}$	$v = \frac{s}{t}$
Force = mass × acceleration	$F = ma$
Acceleration = $\frac{\text{change in velocity}}{\text{time taken}}$	$a = \frac{\Delta v}{\Delta t}$
Momentum = mass × velocity	$p = mv$
Work done = force × distance moved in the direction of the force	Work done = Fs
Power = $\frac{\text{energy transferred}}{\text{time taken}} = \frac{\text{work done}}{\text{time taken}}$ = force × velocity	$P = \frac{Fs}{t} = Fv$
Weight = mass × gravitational field strength	$W = mg$
Kinetic energy = $\frac{1}{2}$ × mass × speed^2	$E_K = \frac{1}{2} mv^2$
Change in potential energy = mass × gravitational field strength × change in height	$\Delta E_p = mg \, \Delta h$
Pressure = $\frac{\text{force}}{\text{area}}$	$P = \frac{F}{A}$
Pressure × volume = number of moles × molar gas constant × absolute temperature	$pV = nRT$
Charge = current × time	$\Delta q = I \, \Delta t$
Potential difference = current × resistance	$V = IR$
Electrical power = potential difference × current	$P = VI$

Potential difference = $\frac{\text{energy transferred}}{\text{charge}}$	$V = \frac{W}{q}$
Resistance = $\frac{\text{resistivity × length}}{\text{cross-sectional area}}$	$R = \frac{\rho l}{A}$
Energy = potential difference × current × time	$E = VIt$
Capacitance = $\frac{\text{charge}}{\text{potential difference}}$	$C = \frac{q}{V}$
Wave speed = frequency × wavelength	$v = f\lambda$
Centripetal force = $\frac{\text{mass × speed}^2}{\text{radius}}$ Force between point charges = $\frac{\text{constant × charge 1 × charge 2}}{\text{(separation of charges)}^2}$	$F = \frac{mv^2}{r}$ $F = \frac{kq_1q_2}{r^2}$ $k = \frac{1}{4\pi\varepsilon}$
Force between point masses = $\frac{\text{gravitational constant × mass 1 × mass 2}}{\text{(separation of masses)}^2}$	$F = \frac{Gm_1m_2}{r^2}$
Transformer equation: $\frac{\text{potential difference across coil 1}}{\text{potential difference across coil 2}} = \frac{\text{number of turns on coil 1}}{\text{number of turns on coil 2}}$	$\frac{V_1}{V_2} = \frac{N_1}{N_2}$

The following formulae may be useful in the study of chemistry. You should check your syllabus to determine which ones are included.

Reaction rate = rate constant × (concentration of reactant A)^m × (concentration of reactant B)^n	Rate = $k\,[A]^m[B]^n$
Arrhenius equation: ln(rate constant) = ln(frequency factor) − $\frac{\text{activation energy}}{\text{molar gas constant} \times \text{absolute temperature}}$	$\ln k = \ln A - \frac{E}{RT}$
Equilibrium constants for the equation: $wA + xB = yC + zD$ Concentration equilibrium constant K_c where [A], [B], [C] and [D] are the concentrations of the reactants and products; w, x, y and z are the numbers of moles Partial pressure equilibrium constant K_p where p_A, p_B, p_C and p_D are the partial pressures of the reactants and products; w, x, y and z are the numbers of moles	$K_c = \frac{[C]^y[D]^z}{[A]^w[B]^x}$ $K_p = \frac{[p_C]^y[p_D]^z}{[p_A]^w[p_B]^x}$
pH = – log (hydrogen ion concentration) pK_a = – log (acid dissociation constant)	pH = –log [H$^+$] pK_a = –log (K_a)

SUMMARY OF STATISTICS RELATIONSHIPS

Mean $\bar{x} = \frac{\Sigma x}{n}$
Standard deviation $s = \sqrt{\frac{\Sigma(x-\bar{x})^2}{n-1}}$ or $\sqrt{\left(\frac{\Sigma x^2}{n} - \bar{x}^2\right)}$
Standard error = $\frac{s}{\sqrt{n}}$
Chi-squared $\chi^2 = \frac{\Sigma(O-E)^2}{E}$
Degrees of freedom = (rows –1) × (columns –1)
$t = \frac{\text{difference between means}}{\text{standard error in difference between means}}$
Standard error in difference = $\sqrt{\left(\frac{s_1^2}{n_1} + \frac{s_2^2}{n_2}\right)}$
Mann–Whitney U_1 value = $n_1 n_2 + \frac{n_2(n_2+n_1)}{2} - \Sigma R_2$ where ΣR_2 is the sum of ranks for data set 2
Spearman rank correlation coefficient $r_S = 1 - \frac{6\Sigma d^2}{(n^3-n)}$
Simpson's diversity index $D = \frac{N(N-1)}{\Sigma n(n-1)}$ where n is the number of all species and N is the total number of individual species present

SOME USEFUL PHYSICAL CONSTANTS

c	Speed of light in vacuo	3.0×10^8 m s^{-1}
e	Charge on an electron	-1.6×10^{-19} C
m_e	Rest mass of an electron	9.1×10^{-31} kg
m_p, m_n	Rest mass of a proton or neutron	1.7×10^{-27} kg
μ_0	Permeability of free space	$4\pi \times 10^{-7}$ H m^{-1}
ε_0	Permittivity of free space	8.9×10^{-12} F m^{-1}
h	Planck constant	6.6×10^{-34} J s

G	Universal gravitational constant	6.7×10^{-11} N m^2 kg^{-2}
N_A	Avogadro constant	6.0×10^{23} mol^{-1}
R	Universal gas constant	8.3 J mol^{-1} K^{-1}
k	Boltzmann constant	1.38×10^{-23} J K^{-1}
F	Faraday constant	9.6×10^4 C mol^{-1}
σ	Stefan's constant	6.7×10^{-8} W m^{-2} K^{-4}

Answers to 'Questions to try'

Page 8

1. (a) 1000 000 (b) 0.001 (c) 0.01 (d) 144
   (e) 4 (f) 0.2 (g) 1 (h) d × d
2. (a) 10^4 (b) 10^{-4} (c) 10^{-3} (d) 10^2
   (e) 10^6 (f) 10^{-3} (g) 10^{-4} (h) 10^{-5}

Page 9

1. (a) 625 (b) 20 (c) 10^9 (d) 10^{18}
   (e) 10^8 (f) m (g) 1 (h) 4
   (i) 27 (j) 10^{-6} (k) 1/8 (0.125) (l) 10
   (m) 10^{-4}
2. (a) Pa^{-2} (b) $kg\ s^{-1}\ K^{-1}$ (c) $mol^{-2}\ dm^6$

Page 10

1. (a) 2.2×10^3, 2.2×10^3 (b) 4.7×10^{-5}, 47×10^{-6}
   (c) 2.5×10^{-3}, 2.5×10^{-3} (d) 1.35×10^8, 135×10^6
   (e) 2.5×10^6, 2.5×10^6 (f) 1.2×10^{-4}, 120×10^{-6}
2. (a) 3300 (b) 0.0000048 (c) 160 000 000
   (d) 27

Page 11

1. 4.64×10^3m 2. 8.5×10^{-27}kg
3. (a) 3.0×10^8 (b) 4.0×10^3 (c) 4.7×10^9
4. $9.1 \times 10^{12}m^2$ 5. 2.2×10^{21}N
6. $1.7 \times 10^{-29}m^3$

Page 12

1. (a) $kg\ m^2\ s^{-3}\ A^{-1}$ (b) $kg\ m^{-1}\ s^{-2}$
   (c) $kg\ m^2\ s^{-3}\ A^{-2}$ (d) $kg\ m^2\ s^{-2}$
2. (a) s^{-2} (b) $kg\ m^2$ (c) $A\ s\ mol^{-1}$
3. (a) yes (b) ..yes (c) ..no (d) no

Page 13

1. (a) 0.0111 m (b) 3.5×10^{-8} s (c) 35 000
   (d) 30.6 $m\ s^{-1}$ (e) 9 180 000 Hz
   (f) $2.24 \times 10^{-3}m^3$ (g) $2.6 \times 10^{-6}m^2$
   (h) $1.5 \times 10^{-3}m^3$ (i) 0.037 s^{-1}
   (j) 1.5×10^{-8}F
2. (a) 2.42×10^5Pa (b) 4.3×10^4 Pa
   (c) 1.8×10^5 Pa (d) 1.33×10^5 Pa
   (e) 1.3×10^3 Pa
3. (a) 7600 $kg\ m^{-3}$ (b) 2.3×10^{-4} $N\ m^{-2}$(Pa)
   (c) 760 $J\ s^{-1}$ (W)

Page 14

1. (a) 5.0 ns (b) 1.5 Mm (c) 20 $GN\ m^{-2}$
   (d) 63 μm^3 (e) 1.5 kW (f) 1.6 ns^{-1}
2. (a) 4 (b) 4 (c) 2 (d) 3
   (e) 5 (f) 5 (g) 3
3. (a) (i) 9.6×10^4 $C\ mol^{-1}$ (ii) 9.65×10^4 $C\ mol^{-1}$
   (iii) 96.49×10^3 $C\ mol^{-1}$
   (b) 96.485309 $kC\ mol^{-1}$ (c) 1×10^{-6}%

Page 15

No answers provided: You should check your order of magnitude calculation against an answer obtained using a calculator.

Page16

1. -2.1×10^{-18} J 2. -4.8×10^{10} J 3. 28.2°

Page 17

1. 12.5:1 2. 1.63 $m\ s^{-2}$ 3. (a) 1/16 (b) ≈22

Page 19

1. 911 W 2. 83 mm 3. 1.5×10^{-3}% 4. C H_2 Cl
5. (a) 1.05 MW (b) 7.2% (c) 14.6 MW

Page 21

(b) $GM/r = v^2$ (e) $m\ s^{-2}$ (f) s^2
(g) $mol^{-1}\ dm^3\ s^{-1}$ (h) 1 (no unit) (i) Pa^{-2}

Page 25

1. (a) $v = \sqrt{(u^2 + 2as)}$ (b) $s = (v^2 - u^2)/2a$
2. (a) $p = nRT/V$ (b) $n = pV/RT$
3. $C = 1/(4\pi^2 Lf^2)$ 4. $(K_a[HA])^{1/2}$
5. (a) $f = (2/\pi)\sin^{-1}0.5$ (b) 19 Hz
6. (a) $t = -35 \ln(2/6)$ (b) 38 s

Page 27

1. 1.86 2. 21.1 3. 8.7×10^{18} 4. 3.7×10^{19}
5. 4.3×10^4 6. 3.04
7. (a) 0.259 (b) 62.9° (c) 0.967
8. 5.2×10^3 9. –0.69 10. 7.7

Page 29

1. (317 ± 2) mm 2. (36.6 ± 1.7)g
3. R = (4.2 ± 0.3) Ω P = (6.5 ± 0.5) W
4. (a) Mean = 1.23 s Estimated SE = 0.01 s
5. g = (9.8 ± 0.2) $m\ s^{-2}$; uncertainty = 2(.1)%;

Page 30

1. Area = 3.5×10^{-3} m^2; perimeter = 0.53 m
2. 4320 mm^2; 4.32×10^{-3} m^2
3. 4.4×10^9 m; 4.0×10^7 m; ratio 110
4. 0.24 mm^2 or 2.4×10^{-7} m^2

Page 31

1. (a) 2.5×10^{-8} m^2 (b) 6.4×10^{-8} m^3 (c) 1.4×10^{-3} m^2
2. 0.12 m^3 3. $3.9 \times 10^{-3}m^2$; $1.64 \times 10^{-5}m^3$
4. 0.062 m
5. (a) 5.1×10^{14} m^2 (b) 2.1×10^{15} m^2
6. 1.15×10^{-44} m^3; 1.5×10^{17} $kg\ m^{-3}$
7. 3.8×10^{-3} m; 1.1×10^{-6} m^2; 4.5×10^{-10} m^3

Page 33

1. 3.4; 3.4 2. 400 × 3. 3.4 s
4. Reduced to 1/24 original flow rate

Page 34

1. (a) 0.62 m^{-1} (b) 4.5×10^{-3} W^{-1} (c) 4.5×10^8 F^{-1}
   (d) 20 s^{-1} (Hz) (e) 6.6×10^{-4} s
   (f) 4.3×10^{-6} s
2. (a) 6.7×10^{-4} m^{-1} (b) 34 m^{-1}
   (c) 1.69×10^6 m^{-1}
3. 1.62 Ω
4. (a) 1.0 m from lens on same side as object (virtual)
   (b) 0.56 m from lens on opposite side to the object (real)

Page 39

1. (a) 4.5×10^{-4} (b) 1.2×10^{-4} (c) 2.3×10^{-4}
   (d) $mol\ dm^{-3}\ s^{-1}$
2. (a) 0.072 $\mu g\ s^{-1}$ (b) 0.17 $\mu g\ s^{-1}$ (c) 0.030 $\mu g\ s^{-1}$

Page 41

1. (a) about 25000 (b) about 40%
2. (a) about 7.5×10^4Pa
3. (a) (ii) acceleration
   (b) (i) work done (ii) distance travelled
   (iii) total change of population (iv) charge
4. (b) about 6500 J

Page 43

1. (a) 0; 25 (b) displacement at $t = 0$
   (c) $m\ s^{-1}$; speed (d) $s = 25t$
2 (a) 26 $kJ\ mol^{-1}$; 0.070 (b) ($\Delta H(vap)$) at 0°C
   (c) $kJ\ mol^{-1}\ K^{-1}$ (d) ($\Delta H(vap)$) = 0.070 θ + 26
3. (a) 0; 4.0×10^{14} (b) F when r is infinite
   (c) $N\ kg^{-1}m^2$ (d) $F = 4.0 \times 10^{14}/r^2$

Page 44

1. B against $1/r$; gradient is μNI
2. T against C; gradient is $0.69R$
3. d against t; gradient is v; intercept is $-s$

Page 45

1. (a) f against $1/l$ (b) $0.5(\sqrt{T/\mu})$
2. (a) $1/\lambda^2$ against E (b) $2m/h^2$

Page 49

1. 1.06×10^{-3} s^{-1}
2. (a) 26.8 h (b) 29.1 h
3. (a) 11 900 years (b) 4.4×10^{20} (c) 7.5×10^9 Bq
4. (a) 60 μA (b) 375 μC

Page 51

(a) 2.6×10^{-9} N (b) 4.8×10^{-10} m

Page 52

1. Data agrees with $f \propto C^{-0.5}$

Page 53

1. (a) 0.57; 0.82; 0.70 (b) 0.69; 0.73; 0.95
   (c) 0.78; 0.62; 1.25
2. (a) 50°; 5.1 N (opposite), 6.1 N
   (b) 60°; 0.52 T (adjacent), 0.60 T
   (c) 20°; 0.69 cm (opposite) 0.73 cm

Page 54

AC = 3.00 cm; BC = 3.35 cm
No answers given for angles.

Page 57

0, 0.106 m, 0.150 m, 0.106 m, 0.057 m, 0, –0.106 m, –0.150 m, 0, 0.106 m

Page 58

1. 85 V
2. (a) 325 V (b) 650 V

Page 59

478 m s^{-1}

Page 60

1. 0.071 N kg^{-1}
2. 21 m s^{-1} approx 3° E of N
3. 40 Ns to S –50 Ns to NE = 83 Ns at 25°W of S

Page 61

1. 3.6×10^{-5}T at 30° to the horizontal
2. 8.7 N
3. 1.3×10^5 m s^{-1} at 92.5° to original direction
4. (a) 2.7 N; 6.8 N (b) 4.6 N; 3.9 N
   (c) 9.5 N; 8.5 N (d) 12.7 N at 42° to horizontal

Page 62

1. (a) 0.79 (b) 0.087 (c) 1.31 (d) 4.71
   (e) 7.85 (f) 436
2. (a) 115° (b) 540° (c) 28.6° (d) $1.7 \times 10^{4\circ}$
   (e) 3.6×10^3
4. (b) 8°

Page 63

1. (a) 15.7 rad s^{-1} (b) 3.9 m s^{-1}
2. 1.25×10^{11} m s^{-2}
3. (a) 48 Hz (b) 0.021 s
4. 105 rad s^{-1}

Page 66

1. log: (a) –0.631 (b) 0.369 (c) 10.369 (d) –5.024
   ln (a) –1.452 (b) 0.850 (c) 28.481 (d) –11.569
2. (a) 1.40 (b) 2.30 (c) 1.00 (d) 12.6
   (e) 13.0

Page 68

1. (b) (15.5±0.2) m^{-1} (c) (45±1) mm
2. n = 0.33 (1/3); Ro = 1.4×10^{-15} m

Page 70

(a) 0.66 (b) 0.48

Page 71

(b) 53 (c) 1.680 – 1.699 m
(d) about 2%

Page 73

(a) 94
(b) mean is 36.7 g; median is 36.5 g; mode is 36.5 g

Page 74

1. Mean = 6.31 mm SD = 0.04 mm
2. 0.41 cm
3. 1.6 g

Page 79

1. $\chi^2 = 1.12$; $\chi^2 < 3.84$ so drug not effective
2. E = mean = 95; $\chi^2 = 7.54$; degrees of freedom = 4
   Critical value = 9.49; $\chi^2 < 9.49$ so no significant difference
3. $\chi^2 = 2.16$; Degrees of freedom = 3; $\chi^2 < 7.82$
   Data consistent with theory

Page 81

$t = 9.54$; degrees of freedom = 5;
critical value = 4.03 at 0.01 level so the difference is significant

Page 83

U values are 10 and 20; Critical value is 3; smallest $U > 3$ so there is no significant difference

Page 85

2. (a) gradient $m = 0.089$ kJ mol^{-1}K^{-1};
   intercept $c = -1.30$ kJ mol^{-1}
   (b) $r = 0.99$ so this is a good correlation
   (c) 59.6 kJ mol^{-1}

Page 87

1. $r_s = +0.588$ The critical value for 9 pairs is 0.68
   $r_s < 0.68$ so there no correlation
2. $r_s = +0.27$ The critical value for 8 pairs is 0.74
   $r_s < 0.74$ so there is no correlation

Page 89

Simpson Diversity Index = 5.28. This is greater than in the example so there is greater diversity.

Page 90

(a) 286 (b) 500

Page 91

Total fish in lake at end = 600
Species A originally = 125
(Species A finally = 100)

Glossary of terms

Confidence limits: the degree to which you can besure that the observations made could not have occurred by chance

Critical values: the value of a quantity derived in a test that tells when particular confidence limit is reached

Cube root: the cube root of a number x is the number that when multiplied by itself twice gives x

Data: the numberical information relating to a particular problem or experimental investigation and from which conclusions are drawn

Degree: 1/360 of the angle subtended at the centre of a circle

Degree of freedom: this is related to the size of a sample. It tells how much the data is free to vary

Denominator: the number below the dividing line in a fraction

Dependent variable: the quantity under investigation. This quantity changes when the independent variable is changed

Efficiency: the ratio of the useful power input of a system

Error bar: lines drawn on a graph that show the uncertainties in the experimental data

Exponent: another term for an index

Exponential change: one in which the quantity always increases or decreases by the same fraction in the same time or in the same distance

Extrapolation: the process of determining a value that is outside the range of experimental data by extending a graph or by calculation

Gradient: the change in the y value of a graph divided by the change in x value (Dy/Dx)

Histogram: a bar chart in which the x axis is divided into continuous class intervals

Hypothesis: an idea that is proposed at the start of an experiment suggesting a possible link between quantities which the experiment sets out to test

Independent variable: the quantity that is controlled in an investigation

Intercept: the place where a graph line cuts one of the axes of the graph

Interpolation: determining a value that is within the range of experimental data by drawing a graph or by calculation

Linear relationship: a graph of one quantity against the other is a straight line which may or may not go through the origin

Magnitude: the size of a quantity

Mean: the sum of a series of quantities divided by the number of quantities added

Median: the middle number in a series. There are as many numbers above it as below it

Mode: the number in a series that occurs most often

Normal distribution: a bell shaped distribution that is a symmetrical about the centre. For a normal distribution the mean and median coincide

Null hypothesis: if a hypothesis suggests no relationship between two quantities the null hypothesis proposes that there is no relationship between them. The analysis of data sets out to test whether this is correct

Numerator: the number above the dividing line in a fraction (fraction = numerator/denominator)

Percentage: a way of expressing a ratio in terms of parts per hundred

Population: the total number of individuals from which a sample is drawn

Prefix: the number placed before a unit to signify the power of 10 by which the number is multipled eg mm = milimetre = 10^{-3} m

Probability: the chance of a particular event occurring

Proportional relationship: a graph for two quantities that are proportional to one another is a straight line through the origin. The ratio of the two quantities is always the same

Radian: the angle subtended at the centre of a circle by an arc equal to the radius

Ratio: the relative sizes of two quantities

Reciprocal: the result of dividing 1 by the number (1/x)

Scalar: a quantity that has only magnitude (size)

SI units: the system of units in which length is in m, mass in kg, time in s, current in A and amount of substance in mol

Significant figures: the number of figures in a number from the first counting from the first non-zero number onwards eg 0.123 and 1.30 are both 3 significant number figures

Sinusoidal: a curve that varies with the same shape as the variation of sin of an angle with angle

Square root: the square root of a number x is the number that when multiplied by itself gives x

Standard deviation: the quantity that is related to the degree of spread of data in a normal distribution

Standard error: the standard error in a set of measurements is the standard deviation divided by the square root of the number of the measurements

Standard form: a number in the form of a numerical part and an exponent part that expressing the powers of 10 relating to the number eg 1.23×10^3

Tangent: a line that touches a curve but does not intersect it

Transposing: changing the subject of a formula eg changing $V = IR$ to $I = V/R$

Uncertainty: the uncertainty expresses the limits within which you are confident that the true value of a measured quantity lies

Variance: this is the square of the standard deviation

Vector: a quantity that has both magnitude and direction